DEDICATION

To my family, who believed I'd get here, even when I didn't.

DOWNLOAD THE AUDIOBOOK FREE!

Thanks for grabbing a copy of my book! As a way to say thank you, I'm giving you the audiobook version completely FREE!

Life is busy, and finding time to sit and read can be tough. That's why I went into the studio and narrated this entire book myself, so it's like we're having a one-on-one conversation about cybersecurity while you're on the go.

Your free audiobook is waiting:

SMCYBER.COM/AUDIO

CYBERSECURITY FOR PEOPLE WHO DON'T HAVE TIME

The Three Pillars for Protecting Your Small Business

SETH BAGDANOV

First Edition 2025

TABLE OF CONTENTS

INTRODUCTION

Most people think of cybersecurity as a way to stop hackers from stealing their data. While that's true, the bigger picture is that cybersecurity is about keeping your business up and running. A cyberattack can shut down operations completely. For a small business, even a few hours of downtime can mean serious financial losses. If the disruption lasts days or weeks, the impact can be devastating.

You might not think about cybersecurity in terms of operational continuity, but you should. Imagine a coffee shop that suddenly can't process credit card payments because of a cyber incident. Customers line up, but the business can only accept cash. Some customers might scramble to find an ATM, but most would just leave. The morning rush—one of the most critical sales periods of the day—is lost.

I've spent over a decade in IT and cybersecurity, helping hundreds of small and mid-sized businesses prepare for, respond to, and recover from breaches. I've led incident response efforts for ransomware attacks and developed recovery plans on the fly. Today, I leverage that frontline experience to provide strategic consulting, helping organizations proactively strengthen their cybersecurity posture.

I'm also ranked among the top performers on advanced cybersecurity challenge platforms—training grounds that test current, real-world problem-solving. I'm certified in both offensive and defensive security. In other words, I don't just know how to set up defenses; I have experience thinking like the people trying to break through them. I understand the tactics attackers use, the mistakes they wait for, and the blind spots they exploit once they're in.

One thing I've seen again and again is how fast critical systems that businesses rely on are suddenly locked, corrupted, or gone altogether. What people take for granted as "just working" vanishes in an instant.

Worse, when systems crash, employees are often stranded. Small businesses rarely have large IT teams standing by to swoop in and save the day. Instead, employees sit idle, waiting for someone to fix the unfixable. Every hour of downtime bleeds money: missed sales, missed deadlines, and missed opportunities. And every delay chips away at customer trust, making the road to recovery steeper and more expensive.

According to a 2024 report from JumpCloud, the average downtime caused by a ransomware attack is twenty-one days.[1] That's three weeks of lost business. Three weeks where a restaurant can't process credit card payments, a law firm can't access case files, a retail store's website stays offline, or a medical office cannot retrieve patient records.

Think about what it would cost if your business had to shut down for even a few days. What about a week? Three weeks? The companies that survive are the ones that prepare before disaster strikes. Companies that build resilience into their operations recover faster, minimize damage, and keep their customers' trust intact.

A strong cybersecurity strategy also helps the business bounce back quickly when something does go wrong. Small businesses don't have to be perfect, but they need to be prepared. The businesses that survive cyberattacks are the ones that expect the unexpected and have a plan in place to deal with disruptions. Because when your business goes down, your customers won't wait. They'll go somewhere else.

In this book, you'll see how strong cybersecurity practices can actually win you more customers and set you apart from competitors. You'll also learn why prevention costs far less than recovery and get a clear picture of just how expensive a cyberattack can be. Most importantly, you'll be guided step by step through building a foundational cybersecurity

---

[1] Kate Lake, "2024 Ransomware Attack Statistics & Trends to Know," *JumpCloud* (blog), October 22, 2024, https://jumpcloud.com/blog/2024-ransomware-attack-statistics-trends-to-know

program that's practical, scalable, and tailored to real-world risks. Whether you're starting from scratch or shoring up existing defenses, this book is your blueprint for getting security right without getting overwhelmed.

And if it ever feels like too much to handle alone, you don't have to. Just like you'd call in an accountant for taxes or a lawyer for contracts, there are cybersecurity consultants who specialize in helping small businesses like yours. You'll learn more about when to bring in outside help later in the book in Chapter 12, but for now, just know that expert guidance is always available when you feel you're in over your head.

Why Customers Care About Security More than Ever

Customers today are more aware of data security than ever before, and their expectations for businesses have shifted. They expect you to safeguard their data, and they're willing to walk away from businesses that cannot meet that expectation. Every major data breach that makes headlines reinforces this awareness, reminding consumers that their personal information is constantly at risk. It's no longer just big corporations under scrutiny; small businesses are being held to the same standard. If customers sense that a company isn't serious about protecting their data, they won't hesitate to take their business elsewhere.

Part of this shift comes from the increasing amount of personal data that businesses collect. Every online purchase, account sign-up, and digital interaction requires customers to hand over information like email addresses, passwords, and payment details. Customers know that if a business neglects security, their data may be compromised, leading to fraud or even identity theft. Businesses that proactively communicate their security efforts can set themselves apart in an environment where consumers are constantly evaluating whether a company is trustworthy.

Today's customers demand more than just quality products or services; they want businesses that respect their privacy and take active steps to protect their data. A company that openly prioritizes cybersecurity by using secure payment systems, offering two-factor authentication, or displaying a clear privacy policy on their website has an edge over competitors that ignore or downplay these concerns. Cybersecurity is now a customer retention strategy.

Companies that openly communicate their commitment to protecting customer data can turn this into a powerful marketing tool. Think about how businesses highlight their best qualities to attract customers. Restaurants promote fresh ingredients, online stores showcase fast shipping, and service providers emphasize great customer support. Cybersecurity should be treated the same way. A business that openly states, "We protect your data with bank-level encryption," or "We secure your accounts with multifactor authentication," is instantly more appealing than a competitor that stays silent on the subject.

Customers are paying attention. They want to know that the businesses they trust with their information are taking security seriously. Those that make security part of their brand identity can gain a competitive edge. When given the choice, customers will always prefer to do business with a company that makes them feel safe.

The Financial Impact

Many business owners assume that if they ever experience a cyberattack, they will only need to fix a few computers, reset some passwords, and then move on. But the reality is far worse. And unlike a broken window, or a stolen piece of inventory, the impact of a cyberattack can linger for months or even years.

Unfortunately, many small businesses underestimate the cascading costs of an attack. It's not just about losing money to hackers through fraudulent transactions or ransom payments. There's also the loss of productivity when employees are locked out of systems, the cost of hiring IT specialists or forensics experts, and the potential for lawsuits or compliance fines if customer information is exposed. And the financial hit doesn't stop there: A cyberattack can impact vendor relationships, disrupt supply chains, and lead to higher cyber insurance premiums, as providers see your business as a greater risk.

The costs don't stop there. A 2023 study by Vercara found that 75 percent of consumers would sever ties with a company in the aftermath of *any* cybersecurity issue.[2] That means if your company experiences an incident, you risk losing three out of four customers. Regaining that trust, or attracting enough new customers to replace the lost ones, can take years.

Over my career in IT, I recognized a pattern in the breaches I was called in to investigate. The biggest damage didn't come from some high-tech hacker. It came from gaps in the basics. Sometimes it was a missing policy. Sometimes it was an employee who didn't know what to watch for. Sometimes it was aging infrastructure that no one had thought about for years. The specific details changed, but the root causes kept showing up in the same three areas: technology, policies, and people. Misalignment in even one of these three areas can cause a minor incident to escalate into a catastrophic business failure.

The harsh truth is that small businesses are the least equipped to absorb unexpected costs or recover quickly. Many businesses have recovered from a bad quarter or an economic recession, but far fewer have survived a major cybersecurity incident. A study by Nationwide Insurance found that the average cyberattack costs small businesses between $15,000 and $25,000 in direct recovery expenses, but the actual long-term impact can grow much higher.[3] Even more concerning, more than half of businesses don't survive the months to come.

When you compare the long-term damage and hidden costs of an incident to the cost of good cybersecurity, it's clear that prevention is always cheaper than recovery.

---

2 "75% of U.S.Consumers Would Stop Purchasing From a Brand If It Suffered a Cyber Incident," *Vercara*, December 18, 2023, https://vercara.digicert.com/news/vercara-research-75-of-u-s-consumers-would-stop-purchasing-from-a-brand-if-it-suffered-a-cyber-incident

3 Graham Shippy, "Cyberattack recovery time and cost much higher than businesses realize," *Nationwide*, September 22, 2022, https://news.nationwide.com/cyberattack-recovery-time-and-cost-much-higher-than-businesses-realize/

The Target Breach Case Study

Big corporations aren't immune to cyberattacks, and when they cannot protect customer data, the consequences are massive. A well-known case is the 2013 data breach at Target, which exposed just how devastating a cybersecurity failure can be. We'll explore how deep the damage went, not just in terms of financial losses but in lost customer trust and long-term business damage.

During the holiday shopping season, hackers infiltrated Target's payment system and stole forty million customer credit and debit card numbers, along with personal information from an additional seventy million shoppers. The breach happened because attackers exploited a security weakness in a third-party HVAC company that had remote access to Target's network. This allowed hackers to install malware on Target's point-of-sale systems, quietly harvesting customer payment details for weeks before anyone noticed.

The financial fallout was staggering. Target spent an estimated $200 million on legal fees, settlements, and security upgrades. But the real damage was lost customer trust. After the breach was made public, many shoppers stopped using their Target credit cards, worried that their data wasn't safe. The company's earnings plunged by 46 percent in the months following the attack, as customers took their business elsewhere.

For a massive retailer like Target, recovery was possible at a tremendous cost. Now, imagine the same attack hitting a small business. Most small companies don't have the financial cushion to absorb a major security failure, nor do they have the resources to manage large-scale reputational damage. If a major retailer like Target lost nearly half of its immediate earnings because of a breach, how would the same cyberattack affect your business?

Would You Skip Fire Insurance?

Most business owners wouldn't dream of skipping fire insurance. Even if they've never had a fire in their building, they understand that the risk is too great to ignore. Fires are unpredictable, and the damage can be devastating. Yet, when it comes to cybersecurity, many small businesses don't take the same approach.

The difference is visibility. A fire is dramatic, immediate, and obvious. When flames engulf a building, there's no question about the destruction. But what about a cyberattack? That can happen quietly, behind the scenes, without a single visible sign for days or weeks. By the time a business realizes what's happened, thieves have stolen data, compromised accounts, or locked down critical systems. Many times, the damage is worse than a fire because it doesn't just destroy infrastructure—it destroys trust.

The irony is that cybersecurity protection often costs less than traditional fire insurance, yet many companies are only willing to pay for the latter. A small business might spend thousands each year on commercial property insurance for fires or flooding but hesitate to invest a fraction of that into cybersecurity tools, employee training, and security consultations—even though the risk of a cyberattack is much greater.

Thinking of cybersecurity as an unavoidable cost is the wrong mindset. It's an investment in business survival. Just like a fire prevention system, it's something you hope you never have to rely on but are still grateful to have. Businesses that take cybersecurity seriously are protecting their future, their customers, and their ability to operate without disruption. The question shouldn't be whether your business can afford cybersecurity. A better question is whether you can afford to go without it.

LAYING YOUR CYBERSECURITY FOUNDATION

CHAPTER 1

INTRODUCING THE CYBERSECURITY TRIFECTA

What would happen if your business suddenly couldn't access any of its systems?

No emails. No files. No customer records. No billing, scheduling, or internal documents. Just a locked screen and a ransom demand.

Most business owners assume they'd find a way to push through. But when your systems are frozen and your data's out of reach, even basic operations come to a standstill—and the clock starts ticking fast.

A recent survey asked business owners how long their company could survive a ransomware attack. Three out of four said they wouldn't make it through the week. Nearly half said they'd be out of business in three days or less.

That's how fast things can unravel.

And it's not just about data. It's about lost time, lost trust, and the real risk of losing everything you've worked for.

But here's the good news: Cybersecurity doesn't have to be overwhelming. It boils down to three key areas you need to understand and defend:

1. People – Your employees and team members are the first line of defense and, unfortunately, often the weakest link. Their daily decisions can either protect or expose your business to threats.

2. Policies – A well-defined set of security rules and best practices ensures that everyone follows the same playbook, reducing human error and improving the response when something goes wrong.
3. Technology – The right tools, when used correctly, help detect vulnerabilities, shield your business from threats, and keep your systems resilient.

Think of cybersecurity like protecting your house. You wouldn't just lock the front door; you'd install a security system, add motion lights, and maybe even get a guard dog. These layers complement each other and contribute to a stronger defense overall. The same layered approach applies to protecting your organization.

This book is written for small and medium-sized business owners, managers, and decision-makers who are concerned about cybersecurity but don't want a technical manual. If you're looking for clear, practical steps to strengthen your organization's security posture and want to understand the risks without getting buried in tech-speak, this book is for you. These are the same principles I use with clients to help them recover from ransomware, close serious gaps before they're exploited, and finally feel confident about their security without blowing up their workflow or budget.

And just so you know, every story you'll read in this book is real. These aren't hypothetical scenarios or exaggerated worst-case predictions. They are real-world cases I've personally handled—situations where small businesses found themselves in serious trouble because of security oversights. I've seen the fallout firsthand: stolen money, lost customers, even companies that had to rebuild from scratch after losing their servers. But I've also seen what works—how smart, simple changes can stop cybercriminals in their tracks.

This book isn't here to turn you into a cybersecurity expert. It's here to give you the essentials in plain English and teach enough to help you make informed decisions and protect your business without getting overwhelmed. Cybersecurity doesn't have to be complex, but it does need to be a priority. By taking proactive steps, you can reduce risk and concentrate on what matters most: running and growing your business.

Let's start by breaking down the three pillars of cybersecurity and why they matter.

Pillar One: People

You can have the best cybersecurity technology in the world, but it won't matter if you don't properly train your people to use it. Most cyberattacks rely on tricking people into opening the door instead of breaking it down.

Hackers know this, which is why they often bypass technology altogether and go straight for the easiest attack: exploiting human error. They use tactics like fake emails with malicious links or attachments and social ruses to trick employees. Even the most tech-savvy staff can be fooled if they aren't trained to recognize the red flags.

To strengthen this first line of defense, you need to create a culture where security is second nature. That starts with regular cybersecurity training. Training shouldn't be a one-time event but an ongoing habit that keeps security top of mind.

Employees also need to feel empowered to pause and verify before taking action on a request. If something feels off, employees should know that taking an extra moment to double-check can mean the difference between safety and disaster.

Mistakes will happen, but the real damage comes when they go unreported. Employees should feel comfortable admitting when they've clicked on a suspicious link or made a security misstep. The sooner an issue is caught, the easier it is to contain, so businesses must remove any stigma around reporting potential threats. If employees fear punishment for making a mistake, they're more likely to hide it.

Finally, leadership must lead by example. If management cuts corners on security, employees will follow suit. When leaders actively follow best practices, prioritize security conversations, and reinforce good habits, employees will follow those cues and make cybersecurity a priority in their daily habits.

While employees are often the weakest link in cybersecurity, they can also become your greatest asset if you equip them with the right knowledge.

Pillar Two: Policies

Imagine running a business where every employee just makes up their own rules. Some use sticky notes to store passwords, others reuse the same login for every account, and a few employees never even log out of their computers. It wouldn't take much for a bad actor to slip in and cause chaos. (If any of these bad habits sound familiar, don't worry; this book is for you!)

Without defined policies, there's no structure, no consistency, and no clear expectations. Even the best security tools and trained employees are useless without clear guidelines. The purpose of policies isn't to add red tape or make employees jump through unnecessary hoops. They exist to keep things simple and standardized, so security isn't left to chance.

Many businesses also assume cybersecurity is about buying the right technology, but policies define how that technology is used. You might have the best internet router installed, but if employees don't follow basic security principles, it's like installing a high-tech alarm system but never requiring employees to arm it when they leave. Without policies, people default to whatever is easiest. And usually, this is not the safest choice. Policies make it clear to everyone what's expected of them.

A good cybersecurity policy isn't a thick, unreadable document that gathers dust in a filing cabinet. It should be clear, practical, and easy to follow through on. Employees shouldn't need a law degree to understand security policies. Straightforward guidelines are far more effective than pages of technical jargon that nobody outside of IT understands. But policies can't just exist on paper; they need to be reinforced through regular communication, training, and reminders, and they must be adaptable. Online threats evolve constantly, and so should the rules that protect you. A good policy is regularly reviewed and improved over time. The most effective policies balance security and usability so that employees can do their jobs efficiently while keeping the business safe.

Some businesses treat creating cybersecurity policies as a burden: a menial task to check a compliance box. But in reality, strong policies make running a business easier, not harder. When employees know exactly how to handle sensitive information, avoid scams, and use security tools properly, they'll be less confused and make fewer mistakes. Clear policies eliminate uncertainty and integrate cybersecurity as a natural part of daily operations. Consider security policies as investments that protect your business, employees, and customers, rather than restrictions.

Pillar Three: Technology

Imagine you've built a solid security culture in your business. Your employees are trained to recognize scams, and your policies ensure everyone follows best practices. That's a great start! But without the right tools in place, even the most security-conscious team is still vulnerable.

Technology plays a crucial role in cybersecurity because it acts as a barrier between your business and cyber threats. But here's the catch: Technology isn't a magic bullet. Many businesses assume that if they install antivirus software or a firewall, they're automatically protected. In reality, technology is only effective when it's properly implemented, maintained, and supported by strong policies and trained people.

Let's go back to the analogy of technology as a home security system with cameras, motion sensors, and smart locks. If you never turn them on, forget to charge their batteries, or ignore their alerts, these components are useless. Similarly, cybersecurity tools need to be actively monitored, updated, and used correctly to be effective.

Technology serves two critical purposes in cybersecurity: prevention and detection. Prevention tools, like firewalls and antivirus software, help block threats before they reach your systems. Detection tools, like monitoring systems, help identify potential attacks before they cause serious damage. But no system is foolproof, which is why technology must also support recovery. In the event of a ransomware attack, for instance, backups can mean the difference between an inconvenience and a devastating loss.

At the end of the day, technology is just a tool. A well-secured business can't rely solely on gadgets. You also need policies that define how that tech is used and employees who use it correctly. That's why the three pillars of cybersecurity work together to create a comprehensive defense strategy.

Wrapping Up

With a solid grasp of the three cybersecurity pillars—People, Policies, and Technology—you've built a foundational understanding. We introduced them in that order because it reflects their true priority: people matter most, policies guide behavior, and technology supports both. But while that's the ideal structure, most businesses build their security in the opposite direction. They start with basic tools, add policies as they grow, and finally invest in training and culture to support the people using them. That's why the rest of this book follows that path, beginning with technology and building outward so each pillar makes sense in context and supports the others.

The next step is to identify the specific threats your business faces and learn how to spot them before they escalate into costly disasters.

CHAPTER 2

CYBER THREATS IN PLAIN LANGUAGE

Cyberattacks aren't just a problem for big corporations with million-dollar IT budgets. In truth, small businesses are often the preferred target because they tend to have weaker defenses. The problem is that most small business leadership doesn't even realize they're at risk until it's too late.

You don't have to be a cybersecurity expert to protect your business, but you do need to understand the biggest threats you're facing. Let's break them down into plain, everyday language: These threats include viruses and malware, ransomware, spam or phishing, insider threats, zero-day vulnerabilities, and misconfigurations. Don't worry, we'll skip all the technical jargon and acronyms as we explain each one.

Before diving deeper, let's clarify a few essential terms: *cybersecurity threat*, *threat actor*, *vulnerability*, *mitigation*, *incident*, and *incident response.*

A *cybersecurity threat* refers to any potential event or action that could harm your business's digital systems or data. Threats can range from obvious attacks, like a hacker trying to access your network, to more deceptive tactics, such as emails designed to trick employees into typing in their password. Even natural disasters, like floods or fires, qualify as cybersecurity threats because they can destroy servers or compromise critical data if you aren't prepared.

Behind every cybersecurity threat is a *threat actor.* These are the individuals or groups responsible for attempting to harm your business. They're often called "hackers" and can be motivated by financial gain or simply a desire to cause problems. Threat actors can also come from inside the company itself in the form of a disgruntled employee. Additionally, automated threats may not involve direct human oversight but can still wreak havoc.

A *vulnerability* is any weakness in your defenses that a threat actor could exploit. Think of it as the digital equivalent of a broken window. These gaps provide attackers with opportunities to gain access or cause damage. Vulnerabilities are the weak points where a cyberattack can take hold. Ignoring them is like leaving your front door unlocked.

Mitigation is the process you take to fix these vulnerabilities and protect your business. It's like locking those doors, fixing the windows, and putting up barriers to keep attackers out. While it's impossible to eliminate every risk, the goal of mitigation is to make it as challenging as possible for attackers to succeed.

A cybersecurity *incident* occurs when a threat actor successfully exploits a vulnerability. This can range from a major event, like a ransomware attack that locks your files and demands payment, to something minor, like a phishing email that arrives in your inbox. Incidents vary in their severity, but even small issues can have significant ripple effects and quickly become larger problems.

Incident response is your business's plan for managing these situations. When an incident occurs, the response process involves identifying the issue, containing it to limit further damage, and restoring normal operations as soon as possible. After the fact, incident response also includes learning from the event to improve your defenses and reduce the likelihood of future attacks. Think of incident response as your digital emergency preparedness plan. It helps your business recover from setbacks while building resilience moving forward.

Now that we have the first vocabulary terms defined, let's discuss our first threat: viruses and malware.

Viruses and Malware

Julie, the accounts receivable specialist at a small manufacturing firm, had been dealing with an annoying problem for weeks. Every time she browsed the web, random ads would pop up. Sometimes they'd be sketchy "free gift card" offers, and others were weird supplements she'd never searched for. At first, she'd just close them out and move on with her day.

Then it got worse.

A couple of times a day, her screen would freeze completely, forcing her to pull the plug on her computer just to reboot it. Only five minutes were wasted each time, but she had invoices to send and payments to process, so she blew it off. *Just a nuisance*, she figured.

But a few days later, she noticed something strange. Customer payments stopped coming in. After calling a few clients, she made a horrifying discovery: Someone had emailed them from her email, instructing them to wire payments to a different bank. Several customers had already followed the fraudulent instructions and sent tens of thousands of dollars to a scammer instead of the company.

Julie hadn't realized it, but her computer was infected with *malware*: malicious software that had not only slowed her computer but also compromised her email account. And now, instead of dealing with a few pop-ups, the entire company was facing a financial disaster.

Many people assume malware is just an annoyance—a few extra ads here, a slower computer there. But in reality, malware can be far more sinister. It can lurk in the background and steal logins, keystrokes, and sensitive data without any obvious warning signs.

Let's break down what malware really is, how it spreads, and why even minor infections can have major consequences. Viruses and malware are malicious software programs designed to disrupt, damage, or gain unauthorized access to computers and networks. These programs can infect a single machine or spread throughout an entire organization. Malware can disguise itself as legitimate software or exploit system vulnerabilities to infect computers unnoticed, making it a constant threat to businesses of all sizes.

This category of threats is one of the broadest and most dynamic in cybersecurity, encompassing a huge variety of software that continuously evolves to outsmart defenses. For small businesses, this constant evolution poses a significant challenge. New forms of malware emerge hourly, employing increasingly sophisticated tactics.

Different types of malware carry out specific kinds of attacks. Viruses attach themselves to legitimate programs or files and spread when the infected file is shared. Worms are standalone programs that self-replicate to any computer they can find without needing a user to take action. Trojans disguise themselves as safe or useful software while secretly carrying out malicious activities, such as opening secret connections to attackers.

Rootkits are a particularly stealthy form of malware. They load before your computer's operating system to hide from antivirus tools and avoid detection entirely. Once installed, a rootkit can give attackers full control over a machine without the user ever knowing it's there. It's like a Hollywood heist movie where the thief sneaks in wearing a staff uniform, walks right past security, and feeds the cameras a loop of empty hallways while they break into the vault. From the security console, everything looks normal. But behind the scenes, the thief has taken over.

Spyware is a category of malware that watches your activities and reports back to a hacker. Keyloggers capture keystrokes to steal sensitive information, like passwords, and stealers extract specific data to sell on the dark web. Each of these poses unique risks to small businesses and requires its own set of detection mechanisms.

The effects of malware on small businesses are often underestimated, partly because they aren't always immediate or obvious. For instance, adware infections that cause pop-up ads might seem like minor annoyances. But when multiple employees experience frequent disruptions, even five-minute delays here or there can quickly add up to hours of lost productivity every day.

While adware might clog up your screen with pop-ups and spyware quietly steals your login credentials, there's one category of malware that takes disruption to a whole new level: ransomware.

Ransomware

One Monday morning, Bill unlocked the warehouse containing the inventory for the online business he worked for, ready to package orders from a busy three-day weekend. His phone had been buzzing off the hook as new order emails came in. As usual, he booted up his computer to check the latest sales reports. But instead of his normal login screen, he was met with an error: "Profile Corrupted."

Confused, he tried again. No luck.

Then he checked the shared folders, only to find all the files renamed with a strange ".locked" extension, including the critical "current-orders.xlsx" spreadsheet. That's when panic set in.

A call to IT confirmed he'd been hit with the worst kind of malware: ransomware. Hackers had encrypted the company's data and were demanding payment to unlock it. Worse, hackers had also compromised the backup system, and the last viable restore point was from Thursday night. Everything from Friday forward was lost. Orders, invoices, shipping confirmations—all gone.

Instead of loading boxes, Bill and his team spent the day scrambling to reconstruct lost data, contact frustrated customers, and untangle a mess that could have been avoided.

While adware might cost small businesses valuable time through lost productivity, ransomware has the power to bring operations to a complete standstill. This makes ransomware one of the most destructive and feared types of malware, warranting a closer look at how it works and its specific implications for small businesses.

Ransomware is malicious software that encrypts a victim's files and renders them inaccessible until a ransom is paid, usually in cryptocurrency. It typically spreads through phishing emails, malicious links, or vulnerabilities in systems. Once it infects a computer or network, ransomware locks down files and displays a message demanding payment for their release. Unlike other malware that might steal or corrupt data, ransomware holds data hostage and forces businesses into a critical decision: pay the ransom or lose everything.

For small businesses, the impact of ransomware can be devastating. Imagine a scenario like Bill's, where all the important files in the business have been locked. This could cause hours, days, or even weeks of downtime, depending on the extent of the infection and the speed of recovery efforts. Delays in recovery, made worse by outdated hardware or insufficient IT support, only add to the frustration and financial burden.

Ransomware is one of the most devastating cyber threats, but how does it actually get into your business in the first place? Hackers don't just randomly break into systems; they often trick someone on the inside into opening the door for them. And one of their favorite ways to do that is through spam and phishing emails.

Spam and Phishing

Kathy ran a small but successful bakery, specializing in custom cakes. Her Gmail address was on every business card, flyer, and social media page, making it easy for customers to find her info. But one morning, she opened her inbox to chaos—1,500 new messages overnight, with more flooding in every minute. Ads for sketchy pharmaceuticals, fake prize winnings, and gibberish-filled emails clogged her inbox. She tried blocking the senders, but each message came from a new email address, making it impossible to stop the flood.

Just like that, her business came to a standstill. Every customer order, every invoice, every new inquiry was buried in the mess. She had to sift through thousands of junk emails just to make sure she didn't miss a real client. While spam overwhelmed Kathy's inbox, the bigger danger was lurking beneath the surface. Many of these messages were phishing emails, designed to trick her into clicking malicious links or handing over login credentials. Spam may seem like an annoyance, but phishing is one of the most prevalent cyber threats for small businesses.

Spam is the digital equivalent of junk mail: unsolicited messages sent in bulk, often with irrelevant or promotional content. While most people are familiar with some spam in their inbox, the problem becomes far more significant when the volume is overwhelming. Imagine waking up to find your inbox flooded with hundreds of spam messages. Suddenly, managing essential communication becomes a nightmare. Wading through irrelevant emails to find a handful of legitimate messages wastes valuable time and undermines productivity.

Phishing, a more targeted and dangerous form of spam, takes this issue a step further. Rather than sending generic messages, phishing emails are crafted to deceive recipients into sharing sensitive information, like passwords or credit card numbers. These emails often impersonate trusted organizations, like banks or service providers. For example, a phishing email might claim to be from your bank, warning you of suspicious activity and urging you to log in and verify your account details. The link, however, leads to a fake website that looks like the bank's but will actually send your login credentials to the threat actor.

The consequences of a successful phishing attack can be devastating, especially for small businesses. With fewer employees handling more responsibilities, a single compromised account can expose a much larger share of sensitive data. Worse, phishing campaigns often target multiple employees at once, multiplying the damage if more than one person falls for the bait.

Phishing attacks come from the outside, but sometimes, the biggest security risks come from within your own business—and they don't always involve bad actors. While we tend to picture hackers as shady figures in dark hoodies lurking in cyberspace, some of the most devastating security breaches happen because of the people inside your company.

Insider Threats

Watson wasn't happy. He'd been written up again last week for making inappropriate comments to the receptionist, and he'd decided this was the last straw. So, when a competitor reached out with a job offer, he jumped at the chance for a fresh start—and a little revenge.

On his last day, he plugged a USB drive into his workstation and copied everything from the company's shared files—customer lists, pricing models, vendor agreements, even upcoming project plans. He slipped the drive into his pocket and walked out the door for the last time, with no one the wiser.

A month later, his former employer was bleeding clients. The competitor had slashed prices and was undercutting them at every turn. Loyal customers were suddenly switching over, and the damage was already done.

Insider threats come from individuals within the organization: disgruntled employees, careless contractors, or even well-intentioned staff who accidentally expose sensitive information. Unlike external attackers, insiders are already trusted with a level of access to systems and data, making them uniquely positioned to cause harm. Whether driven by malice, negligence, or financial gain, employees with access to inside information pose one of the biggest security risks to any business. Regardless of motivation, insider threats can bypass even the strongest external defenses and create chaos for a business's operations, finances, and reputation.

Small businesses are especially vulnerable to insider threats because of their structure. In smaller teams, roles often overlap, and individuals may have broader access to sensitive information than they would in larger organizations with stricter access controls. For example, consider a small accounting firm with ten CPAs. If one of them acts maliciously, they could easily steal the entire client database of tax information.

Even with strong internal security policies, your business is only as safe as the software and systems you rely on. Unfortunately, sometimes those very tools have hidden weaknesses that no one, not even the developers, knows about. This is where zero-day vulnerabilities come into play.

Patching and Zero-Day Vulnerabilities

Dave was busy and impatient, so whenever his computer nagged him with the dreaded "Install Updates Now?" message, he would click

"Remind Me Later." Updates took forever, and he didn't have time to deal with them, especially when he was in the middle of checking emails, updating spreadsheets, or browsing listings for the next car he wanted to buy. Besides, everything worked fine, so what was the rush?

Late one afternoon, Dave pulled up the payroll spreadsheet he had opened that morning, but instead of numbers and names, he got an error: "This file doesn't exist."

Confused, he called IT to find out what was going on. The answer the tech gave him hit like a punch to the gut: The server had been hit with ransomware. The attackers had exploited a known vulnerability in Windows and the version of Chrome installed on Dave's computer. The flaw had allowed them to run commands on his system just by displaying a malicious ad in his browser—no clicks required. When Dave visited a website that served one of those ads, the ransomware quietly launched in the background. From there, the infection spread to the servers and encrypted payroll files, invoices, and customer records. All of these issues could be traced back to Dave clicking "Remind Me Later" one too many times.

Some dangers don't come from outright neglect or malice; they come from small, seemingly harmless choices. Delaying updates might feel like a minor inconvenience, but in Dave's case, that inaction opened the door to a full-blown ransomware attack.

Unfortunately, skipping or delaying updates on workstations and laptops is a common mistake. These updates—often referred to as patches—are released to fix known security flaws in operating systems and software. While it might seem harmless to postpone an update to a more convenient time, this kind of procrastination leaves systems exposed to risks that could have been easily prevented. Regular patching is one of the simplest and most effective ways to protect your devices.

In contrast to known security holes, a zero-day vulnerability is a flaw in software or hardware that developers or the public are unaware of. Imagine locking all your doors and windows at night, only to find out in the morning that your front door lock had a hidden design flaw you didn't know about—but the burglars who ransacked your house did. Cybercriminals who discover these vulnerabilities first can exploit them before a fix is created, hence the term "zero-day"—there's no time for the developer to prepare. These exploits can lead to widespread attacks in a very short period of time.

Security patches help fix known vulnerabilities, but what about the security settings you control? This is where misconfigurations come into play. Even with the best software and hardware, a single overlooked setting, like leaving a default password in place, can create an open door for hackers. It's like installing a top-of-the-line security system for your home but forgetting to change the default passcode from "1234."

Misconfigurations

Kevin was a managing partner at a CPA firm and had been dealing with a sluggish database for days. It was year-end crunch time, and waiting on slow software wasn't an option. Finally, after a weekend of frustration, he put in a ticket to IT to get things running faster.

But when IT tried to log in to the database server, they hit a roadblock. The admin password had been changed. Nobody in the company had done it. Digging deeper, IT found server logs showing unauthorized access from a location in North Korea. A hacker had gained full control of the database through a forgotten remote access portal that still had its default password enabled.

Kevin had to hire a digital forensics firm to investigate the fallout of this breach, and their findings were devastating: Every client tax return the firm had ever stored was for sale on the dark web. Social Security numbers, financial statements, personal details—everything was compromised. All because of one overlooked setting.

Misconfigurations, like overly broad permissions, default passwords, and exposed databases, are some of the easiest ways for hackers to gain access to sensitive data. You'd think changing these simple settings would be the first lesson in any company's cybersecurity handbook. And yet, almost every week, you hear a story in the news about another database breach.

One of the most common missteps involves user permissions. Over time, employees tend to accumulate access as they take on new projects or temporarily fill in for other roles. But that access is rarely ever rolled back. Eventually, someone ends up with far more access than they need, and that becomes a serious liability. If their account is ever compromised, an attacker suddenly has keys to far more than just that employee's job function.

Default credentials also contribute to misconfigurations. Most devices, from printers to network hardware, come with preset usernames and passwords that are well known and easily found online. If these aren't changed, anyone could access the network and make disruptive changes, or even escalate their access to more sensitive systems.

Improper configuration can leave even fully updated software vulnerable. While security updates correct flaws in the software's code, misconfigurations happen when the software or device is set up using insecure or inappropriate settings. In these cases, the code may function exactly as designed, but the configuration creates an opening for attackers.

These kinds of misconfigurations are entirely avoidable, yet they remain alarmingly common. And the larger or more fast-paced your business becomes, the easier it is for small oversights to slip through. Kevin's team had applied all the latest security updates, but one critical system remained exposed because of a single overlooked setting. Misconfigurations might seem like minor oversights, but they can create another weak link that jeopardizes the entire organization.

Wrapping Up

Most cyberattacks don't start with advanced tactics—they start with simple oversights. As we've seen, many of the most common threats are easy to overlook but just as easy to prevent with the right mindset. For small businesses, spotting and fixing these low-hanging fruit vulnerabilities can make a big difference. It's almost always cheaper and far less painful to prevent an attack than to recover from one.

No system is bulletproof, but addressing the basics puts you ahead of most targets. Awareness is your first line of defense. In the next chapter, we'll explore why so many cybersecurity efforts fall short, and how your business can avoid the same mistakes.

Next Steps

- ☐ Start a Threat Scrapbook shared folder or Teams/Slack channel and start saving screenshots of phishing emails or news headlines about breaches. Review one example in each staff meeting to keep awareness fresh.

Resources:

The Plain-Language Cyber Glossary breaks down the top essential cybersecurity terms in clear, nontechnical language. Use it to boost team confidence and kick off smarter security conversations.

Download it from smcyber.com/glossary.

CHAPTER 3

WHY CYBERSECURITY FAILS AND HOW HACKERS EXPLOIT IT

Frankie "The Brain" Marino had spent months planning the perfect heist. His target? The grand vault of the biggest casino in town. Millions in cash, high-roller chips, and gold bars sat behind layers of state-of-the-art security—motion sensors, biometric locks, armed guards, and an eye-in-the-sky surveillance team watching every inch of the casino. If he could pull off the heist, he'd be set for life.

But that was the big "if."

One mistake, one unexpected hiccup, and he'd be looking at life behind bars. The risk was enormous. Casinos invest heavily in security, knowing that crooks like Frankie might try their luck. Even if he managed to crack the vault, escape would be a nightmare. There'd be an FBI task force, news coverage, and a manhunt chasing his crew across the desert.

So Frankie—always the big brain—took a step back and reevaluated. Why score one big heist when he could hit a dozen smaller ones with far less risk?

Just down the highway, in the heart of a little town, there were small businesses everywhere: a bakery, a mom-and-pop hardware store, and a car dealership with a security system from the 1980s. Each one kept just enough cash on hand to be worth a visit, and their defenses were practically nonexistent. No security cameras, no armed guards, no million-dollar surveillance teams to watch his every move. The sheriff was already stretched keeping the big-town casinos safe, so a string of break-ins at local shops would barely make the news.

Instead of one high-risk job, Frankie and his crew could rob a dozen small businesses in a single night and walk away with a respectable haul. And best of all? No one would track them down afterward. It was simple math: less risk, more reward.

Cybercriminals think the same way.

Breaches happen because of false assumptions that leave companies wide open to attack. Many small businesses believe they're too insignificant to attract hackers. Others assume their IT provider is handling security, never questioning what protections are actually in place. Some invest in buzzword technology like firewalls, antivirus software, and cloud backups, feeling confident they've done enough. But hackers don't need sophisticated tools to break in; they just need one overlooked weakness, one untrained employee, or one outdated system to slip through.

This chapter breaks down the false sense of security that leads to cybersecurity failures and shows how hackers exploit those blind spots. We'll examine why businesses unknowingly make themselves attractive targets, how cybercriminals choose which businesses to attack, and what weaknesses they look for. Along the way, you'll also learn about the most common hacking tactics and why they work so well, and most importantly, what businesses can do to break out of the "easy target" category.

The Illusion of Security

In the hypothetical heist story, Frankie & Co. didn't hit the casino because they knew it would be too much work. They wanted easy targets—businesses with weak locks, no security cameras, and no one watching the side door. Cybercriminals operate with the same mindset, and yet, most small businesses don't believe they're even on a hacker's radar.

It's a common misconception: "We're too small to be a target."

Management assumes hackers are chasing million-dollar paydays, breaking into massive corporations with deep pockets. The truth is far less dramatic. Cybercriminals aren't looking for the biggest payout; they're looking for the *easiest* one. They'd rather hit a hundred small businesses in a day than spend weeks trying to breach a Fortune 500 company's security.

This belief that one is "too small to be a target" does more than just invite hackers; it turns small businesses into ideal targets. Big corporations invest heavily in security by hiring cybersecurity teams, running vulnerability scans, and monitoring their networks 24/7. Small businesses, on the other hand, often run on outdated systems, use weak passwords, and rely on a basic IT provider to "manage security." From a hacker's perspective, this is the digital equivalent of an unlocked back door with a neon *welcome* sign.

Take Green Ford Sales, an Abilene, Kansas, car dealership that thought cybercrime was something that only happened to bigger businesses. One day, without anyone noticing, hackers infected the controller's computer with stealer malware. With stolen passwords, they quietly added nine fake employees to the books. Over the next several months, those fake workers collected over $63,000 in payroll payments before anyone realized something was off. By the time the dealership discovered the fraud and tried to reverse the transactions, $23,000 was gone for good. No alarms had gone off. No IT technician had come to their rescue. The hackers didn't have to be sophisticated criminals; they just needed a target that wasn't paying attention.

But Green Ford Sales isn't a unique case. Hackers know small businesses operate under this false sense of security, and they take full advantage. Automated bots scan the internet constantly, looking for unpatched software, weak passwords, and exposed remote access systems. Once a vulnerability is found, an attacker can quietly slip in, install malware, and steal customer data or sell access to another cybercriminal, all without the business realizing anything has happened.

Small businesses also assume that because they haven't been hacked before, they must be secure. This is like assuming you don't need insurance because you've never been in a car accident. Just because an attack hasn't happened yet doesn't mean it never will. It could simply mean you haven't been targeted yet. Hackers don't care if a business is "important"; they only care if it's vulnerable.

A business that believes it isn't a target likely won't invest in cybersecurity. It won't train its employees. It won't monitor for threats. It will keep running on blind trust, assuming its digital world is safe because it hasn't been burned—until one day, it is.

Overconfidence in IT Departments and Security Tools

If small businesses aren't relying on the "we're too small to be a target" assumption, they often fall into another dangerous trap: assuming that their IT provider has security covered. Business owners believe that paying IT staff to manage computers and networks automatically includes cybersecurity. This applies to both an in-house IT department and outsourced third-party IT service providers. In reality, almost no IT providers are cybersecurity specialists. They are technicians focused on keeping systems running, not actively developing and implementing a cybersecurity program.

IT is like a general doctor—great for routine care, but you need a specialist for major issues. The IT team's job is to set up new laptops, fix network issues, and ensure employees can access their files and applications. They might manage antivirus software, but that doesn't mean they know their way around cybersecurity solutions.

Many business owners don't realize this gap exists until something goes horribly wrong. Take Cameron, the controller at a construction company. He had just fired his old IT provider and switched to a new managed service provider who promised an all-inclusive management package. Their contract included multifactor authentication, EDR software, cloud backups, and real-time device monitoring. At the same price as the old service provider, it sounded like a great deal. Cameron figured that with all these tools in place, there wasn't much left to worry about.

Less than six months later, that confidence came crashing down.

A single employee clicked on a bad link, a virus slipped past their defenses, and their bank account was compromised. The $1.5 million deposit that was supposed to cover next month's wages and fuel disappeared overnight. All of a sudden, Cameron wasn't thinking about security tools. He was desperately trying to claw back enough of the transfer to keep his company running.

What went wrong? Cameron had overestimated the abilities of security tools and assumed they were enough to stop every threat. But security isn't just about having the right tools; it's about using them correctly. The firewall didn't stop an employee from clicking a phishing email. Antivirus software couldn't block an advanced form of malware. And while the IT provider claimed to offer real-time monitoring, it was either misconfigured, neglected, or simply not designed to catch this kind of threat.

Hackers love small businesses that assume their IT provider is handling security. The business sees "cybersecurity" in the marketing brochure and checks the mental box, never asking what that actually includes. They don't review their security settings, don't test their backup recovery process, and don't ask for a documented cybersecurity strategy. They trust that someone else is handling it, and that trust leaves them blind. When an attack hits, they find out too late that what they thought was protection was really just a handful of default tools, loosely managed and never tailored to their business.

Real cybersecurity programs go far beyond antivirus and automated patching. They include ongoing risk assessments, policy development, employee security training, vulnerability scanning, and 24/7 threat detection through a monitored security operations center (SOC). This kind of monitoring is very different from the "real-time monitoring" most IT providers advertise. When an MSP says they provide real-time monitoring, what they usually mean is that their remote management platform (RMM) watches for hardware issues, software updates, and basic system errors—not active cybersecurity threats. A true SOC, by contrast, monitors security events in real time, correlates alerts across systems, detects suspicious behavior, and has security analysts who investigate and respond immediately. It's a completely different type of protection.

Most IT providers aren't equipped to deliver this level of coverage because it takes more than just installing tools; it requires a security-first mindset, specialized expertise, and a team dedicated to defending your business around the clock. And most small businesses don't have the internal staff to manage it themselves, either.

The bottom line? Your IT provider keeps your systems running. That doesn't mean they're also keeping you safe. When "cybersecurity" shows up in their brochure, it usually refers to the default tools bundled into their management platform: basic antivirus, automatic updates, and maybe resource monitoring. If your business isn't asking hard questions and demanding specifics, you're not getting a real cybersecurity program—just a marketing buzzword.

Human Error: The Hacker's Best Friend

Hackers love human mistakes. They don't need to break through a company's firewall if they can simply trick an employee into handing over their VPN credentials. They don't need to crack a complex security system if someone simply clicks the wrong link. For all the advanced tools and security software businesses invest in, the easiest way in is almost always through the people inside the organization.

Most business owners assume their employees "know better" than to fall for scams. They believe employees will recognize a phishing email as a scam and delete it. But the reality is very different. Employees aren't falling for scams because they're careless or unintelligent; they're falling for them because hackers are experts at psychological manipulation.

Take Brian, an employee at a small, family-run agriculture supply company. One morning, he received a notification that the company's email security system had quarantined an incoming message from their fertilizer vendor. Quarantine is a safety feature that automatically holds suspicious emails before they reach the inbox, just in case they contain something harmful. Since the company had worked with this vendor for years, Brian assumed it was a false alarm. He contacted IT and asked them to release the message. A few minutes later, it was in his inbox.

The email explained that the vendor had suffered a security breach and had to switch to a new payment system. To continue placing orders, the farm needed to update its billing details. Naturally, Brian clicked the link, logged in with his vendor credentials, and reentered the company credit card.

By the next morning, the company card was maxed out with fraudulent charges. The email wasn't from their vendor—it was a spoofed phishing email that led to a website designed to steal payment details. The spam filter had flagged it correctly, but IT released it without verifying its legitimacy. Brian, seeing what appeared to be a normal request from a trusted source, walked right into the trap.

This wasn't a case of ignorance; Brian knew phishing emails existed. But the attack worked because it felt routine, came from a familiar name, and created a sense of urgency. Hackers know how to craft messages that catch people off guard and make them act before thinking.

Lauren, a partner at a small law firm, fell for a similar scam. One afternoon, she received a call from "Microsoft Support." The man on the other end sounded professional and claimed they had detected a security issue on her laptop. Having recently purchased the laptop from a major retailer, she was confused at first, but the caller reassured her that this was a known problem with a specific batch of devices. He even referenced the exact store where she had bought it, adding to the illusion of legitimacy.

He directed her to a website that looked exactly like Microsoft's official page and asked her to download a remote help tool so they could "fix" the issue. Seconds later, her screen locked up. The hacker was inside her system, browsing through confidential client folders and downloading sensitive legal documents.

Until that moment, Lauren had never heard of social engineering beyond the classic "Nigerian prince" scam. She had no idea that caller ID could be spoofed or that Microsoft never calls customers directly for security issues. If she had received basic security awareness training, she would have recognized the red flags.

Hackers count on the fact that most employees in small businesses haven't been trained to recognize scams. Many business owners assume their team will "just know" when something feels off—but you can't expect people to spot a threat they've never been taught to look for. (In Chapter 10, we'll dig into how to train your staff to recognize warning signs, avoid common traps, and become your business's first line of defense.)

Human error is the single most common reason cyberattacks succeed, and hackers exploit it in countless ways. Some create emails that look like urgent requests from the company's CEO, instructing employees to process a wire transfer immediately. Others impersonate vendors, tricking employees into updating payment details that send money directly into a hacker's account. Some even use AI-generated deepfake voices to impersonate executives over the phone, convincing employees that the request is legitimate.

Hackers don't need to break into a business's network if they can simply ask an employee to open the door for them. The best security tools in the world won't stop an employee from falling for a well-crafted phishing attack. Firewalls and antivirus software can't prevent someone from willingly giving away their password. And policies like "Don't open phishing emails" mean nothing if employees don't know how to recognize those emails.

Businesses can't eliminate human mistakes altogether, but they can reduce them by building a culture where employees slow down, verify requests, and think critically before acting. Encouraging employees to pause and double-check before responding to urgent emails, resetting passwords, or processing financial transactions can be the difference between business as usual and catastrophic data loss.

The biggest cybersecurity failures don't come from brilliant hackers using cutting-edge tools. They come from simple, preventable mistakes made by employees who were never given the training to recognize threats. If small businesses want to stop being easy targets, they need to start by fixing the biggest security flaw of all: the assumption that people won't make mistakes.

How Hackers Choose Their Targets

Just like burglars case neighborhoods looking for dark houses with unlocked doors, cybercriminals actively search for businesses that make their job easier. The difference? Instead of strolling down a street, they use automated tools to scan anything touching the internet.

Before they even launch an attack, hackers gather information. Every company has a digital footprint, whether or not they realize it. Attackers scrape company websites, LinkedIn profiles, social media, and public records to gather intelligence about their targets. They identify what software a business is using, what email formats employees use, and whether the company has been included in past data breaches. If they find enough useful details, they don't even need to rely on automation. They can craft highly targeted attacks designed to manipulate employees into handing over access.

Even when a business believes its systems are secure, attackers might already have a way in. Stolen credentials are widely available on the dark web, where hackers buy and sell login details exposed in past data breaches. If an employee has been using the same password for years, chances are it's already been compromised. Attackers compile these leaked passwords and test them on every account they can find. Since many people reuse passwords across multiple accounts, this tactic is alarmingly successful.

A business with unpatched software, a remote access portal without multifactor authentication, or a router still using default credentials is an easy target. To an attacker, finding one of these is like finding a house with the windows open and the alarm turned off. They don't need a detailed profile of the business or an elaborate plan. They just need one foothold.

Once a foothold is established, the real attack begins. Hackers chain vulnerabilities together and move step by step toward their ultimate goal:

stealing data, deploying ransomware, or taking control of critical systems. It's a lot like how Frankie "The Brain" approached his break-ins.

Frankie didn't smash a window and hope for the best. He slipped through a broken gate after hours, circled to a service door left open for the night cleaning crew, and walked in, carrying a clipboard like he belonged there. At the front desk, he found a master key tucked in a drawer, and a few doors later, he was inside the manager's office, staring at an unlocked computer with the safe code saved right on the desktop. Frankie cracked the safe, emptied the cash, and grabbed client lists and anything else he could resell to a shady competitor. By the time the staff arrived the next morning, he was long gone—and so was half the business's future.

Just like Frankie worked in steps, hackers also work in progression. A typical breach follows predictable stages:

- Initial access – finding a way in using stolen credentials, unpatched systems, or exposed remote access.
- Escalation – stealing more credentials, installing malware, or exploiting internal systems to gain higher access.
- Lateral movement – exploring the network, quietly mapping out where sensitive data and critical systems live.
- Execution – launching the final blow: encrypting files, stealing financial data, or setting up long-term backdoors.

That's why professional defenders don't wait for obvious signs of trouble. They study the stages of attacks using models like the Cyber Kill Chain (developed by Lockheed Martin) or the MITRE ATT&CK framework. These models map each stage of an attack, helping defenders recognize how hackers operate and where they can be stopped. Trained SOC analysts use these frameworks to spot the early indicators of an attack, like unusual login behavior or data moving where it shouldn't.

The earlier in this chain a breach is detected, the easier it is to stop. If no one is watching—or if the business assumes its IT provider is watching—attackers can move freely for days, weeks, or even months before striking.

Most small businesses never realize they've been breached until the final "Actions on Objectives" stage, when ransomware locks their files, customer data is leaked online, or their bank accounts are drained. By then, the damage is already done, and the attack is over.

That's exactly what happened to a small logistics company that relied on remote desktop connections over the public internet. Their IT team had failed to patch a known remote desktop vulnerability, and a ransomware gang discovered the exposed system. Once inside, they encrypted every company server and demanded payment for the decryption key. The business was forced to shut down its digital operations entirely, scrambling to switch back to paper-based tracking while they tried to recover. What should have been a routine Windows update had turned into a crisis that cost the company tens of thousands of dollars for a ransom payment, plus over a week of downtime.

Most businesses never even realize they've been marked as a target until it's too late. Hackers aren't looking for a challenge—they're looking for the easiest way in. They aren't spending hours trying to break through military-grade firewalls or designing complex malware to bypass state-of-the-art security systems. Instead, they follow a much simpler strategy: find the weakest link and exploit it.

Multilayered Protection

Picture a medieval castle standing high on a hill. It has towering walls, a deep moat, armed guards, and a reinforced gate. No single defense is enough to keep intruders out, but together, these barriers create a formidable stronghold. Attackers might breach one wall, but they still have to face the next layer of defense.

Many businesses approach cybersecurity by installing antivirus software or enforcing strong passwords but fail to see the bigger picture. That's the equivalent of dumping the crown jewels in a chest on the hilltop and building one wall around it. A single layer of protection is never enough.

Cybersecurity is not a single product; it's a strategy. And like any good strategy, it requires layers of protection that work together to reduce risk, detect threats early, and minimize damage when an attack inevitably happens. This layered defense approach helps keep your business running smoothly, no matter what. Businesses that treat security as an ongoing strategy rather than a last-minute patch are the ones that thrive, even in the face of attacks.

In the following chapters, we're going to shift gears from understanding what could go wrong to learning how to actively protect your business. The goal isn't to make your business impossible to hack. Instead, the aim is to make your business too difficult a target for cybercriminals to bother with. And when an attack does occur, you want to be in a position where you can recover quickly, minimize damage, and keep serving your customers. In today's world, properly implemented cybersecurity becomes a business advantage, a financial safeguard, and a necessity for long-term success.

Beyond the Three Pillars: Expanding Your Cybersecurity Strategy

The three pillars of cybersecurity—People, Policies, and Technology—give you a strong foundation. But a foundation isn't a finished structure. It's a starting point. To build a real, working cybersecurity program, you need to layer on additional support. These are practical systems that bring the pillars to life and keep them standing strong over time.

That's where these six domains come in. Each domain focuses on a specific area where small businesses tend to fall short, from device protection to disaster recovery. Together, they transform your cybersecurity from a checklist into a layered system you can manage and maintain. Think of them as the framing, walls, and roof of your security house: the structure that turns your foundation into something usable and secure.

These domains give you practical focus areas that translate big-picture strategy into real-world protection:

1. Physical Security – Cybersecurity doesn't just live in the cloud or on your computer; it starts in the real world. If someone can walk into your office and grab a laptop, your digital defenses won't matter much. This domain is about making sure your physical space isn't an open door for security risks.
2. Endpoint Protection – Every laptop, phone, or server in your business is a doorway. This domain is about making sure those devices are protected, watched over, and kept in check so they don't become an easy way in.
3. Network Protection – Think of your network like a digital highway connecting all the parts of your business. Without guardrails, anyone can cruise through. This domain is about keeping your traffic organized, safe, and under control so threats don't spread or slip through unnoticed.
4. Data Access Controls – Not everyone in your company needs access to everything. This domain is about being intentional with who can see what. When sensitive info is left wide open, it's only a matter of time before someone grabs something that they shouldn't.
5. Disaster Recovery – Things break. Systems fail. And sometimes, attackers still get through. This domain is about having a solid plan for getting back on your feet. Backups, recovery steps, and knowing what to do when things go sideways can make all the difference.
6. Human Resilience – People are at the heart of every business and every breach. But they're also your best line of defense when they're trained and prepared. This domain is all about helping your team spot danger, make smart choices, and respond when something feels off.

Each of these six layers supports and reinforces the others, creating a comprehensive security framework that makes your business significantly harder to attack. Cybercriminals thrive on weak links, but this multilayered approach ensures that no single failure point leaves you vulnerable.

Wrapping Up

We've seen how common misconceptions and everyday habits can quietly leave businesses exposed. When attackers look for easy targets, it's often these overlooked gaps that make the difference. Too often, companies don't act until after something goes wrong.

But now, with a better understanding of these pitfalls, you can start building a stronger defense before trouble hits. In the next section, we'll take the three pillars you've learned and apply them across six key areas: Physical Security, Endpoint Protection, Network Protection, Data Access Controls, Disaster Recovery, and Human Resilience. This layered, practical approach will help you reduce risk, bounce back faster, and stay in control when challenges come your way.

Next Steps

- ☐ Identify one "easy target" area in your business that a hacker might exploit (e.g., an old system, a shared password).
- ☐ Adopt a quick "pause and verify" rule: Before acting on any urgent or high-authority request, confirm it by phone or in person.

Resources:

Grab the "Decoding Deception" cheat sheet: a one-page explainer of the three tricks hackers love most. Perfect for posting, sharing, or kicking off your next team meeting. Go to smcyber.com/deception.

CHAPTER 4

ASSESSING AND DESIGNING A RESILIENT SECURITY STRATEGY

Have you ever had a fire marshal show up at your office for a routine inspection? It's usually no big deal. But as they walk through, things may start to go sideways. The emergency exit? Blocked by a stack of boxes that no one's moved in years. The fire extinguisher? Expired sometime before the last Winter Olympics. And the smoke alarms? The batteries were cannibalized to keep the office stereo remote working.

You thought you were prepared until someone actually checked.

Now, swap out the fire marshal for a hacker. Instead of fire hazards, they're looking for security hazards, like outdated software and unprotected access points. And, just like the workplace inspection, you don't really know what's at risk until you take a good, hard look at your cybersecurity.

That's what this chapter is about: performing your own cybersecurity audit before an attacker does it for you. We'll go step by step, identifying your critical assets and checking for weak points. By the end of this assessment, you'll know exactly what's at risk and what to do about it.

As you work through this chapter, use the checklist from smcyber.com/assessment. This list will help you track your progress, document key findings, and pinpoint areas that need immediate attention. A well-documented assessment makes it easier to prioritize improvements and build a cybersecurity strategy that fits your business's unique needs.

It's worth noting that identifying gaps in your security isn't a failure, but a necessary first step toward strengthening your defenses. Every business, no matter how prepared, has areas that can be improved. The key is thoroughness. A rushed review won't reveal the real risks lurking beneath the surface. Instead, take the time to assess every aspect in this chapter, from password policies and access controls to data backups and employee training. Think of this as a diagnostic checkup for your business: You're catching minor issues before they become major security incidents. A strong understanding of existing vulnerabilities allows for more targeted and effective cybersecurity improvements.

Technology Assessment

Identify and inventory critical assets

Every business, no matter how small, has digital and physical assets that keep it running. Some are obvious, like customer records and financial data. Others are easy to overlook until you need to do without them, like the workstation that holds all your QuickBooks files or the Wi-Fi access point that everyone connects to. These assets are the backbone of your operations. If they are lost, stolen, or compromised, your business could not function normally. Think of these as the crown jewels of your business: If a thief broke in, what would they want to steal first?

Start by considering the types of data your business handles. Customer information is usually the most valuable target for cybercriminals. Financial records, such as payroll, invoices, tax documents, and banking credentials, are just as critical.

Beyond data, think about your business's intellectual property. For some companies, this might mean trade secrets, proprietary software, or research data. For others, it could be as simple as a client database, internal workflows, or an extensive email history that holds key business insights and contacts. Even if your business doesn't develop unique products or software, losing these materials could mean starting from scratch.

Once you've identified data assets, shift your focus to the physical hardware that supports your operations. Consider every device that stores or processes business information. Your workstations might hold key documents, email access, and saved credentials. If you have a central server, it likely stores shared files or databases. If you use a point-of-sale system, losing it to a cyberattack could prevent you from processing transactions entirely. Your essential network equipment—like your modem and router—are often overlooked but are the linchpin to keeping your business connected. These are already included by default on the checklist because of their importance.

To identify your assets, start by listing everything you cannot afford to lose in an incident. If your computer crashed tomorrow, what data would be irreplaceable? If someone hacked into your email, what accounts and services would they gain access to? Think about the worst-case scenario that could ruin your business and work backward from there. *List all of these assets on the checklist.*

Then go through and assign them into one of the following priority tiers:

- Core Tier assets are absolutely mission-critical. If they go down, your business stops. Think: customer databases, billing platforms, internet connectivity, or your primary email account.
- Operational Tier assets help you deliver services efficiently. You could survive without them temporarily, but it would slow you down or require manual workarounds, like a diner operating cash-only.
- Support Tier assets are low-impact tools or systems that would be inconvenient to lose but wouldn't stop business operations. For example, if one of several checkout stations at a retail store went down, it might slow things down, but you could continue serving customers.

Document where assets are physically located

Once you've identified and prioritized your critical assets, the next step is to document their physical locations. Physical location plays a key role in determining the specific protections and measures required to safeguard these assets. Creating an inventory of physical locations allows you to identify potential security gaps, such as devices left in unsecured areas or those accessible to unauthorized individuals.

Start by mapping out the locations of all physical systems involved in your operations. This includes desktops and laptops, which may be distributed across multiple floors, branch offices, or employees' homes. Servers are often housed in a dedicated server room or IT closet but could also be in less-protected spaces, such as an unlocked office. Internet modems and network devices, like routers and switches, are equally important.

Mobile devices, such as work phones and tablets, should also be considered. These devices frequently leave the office, increasing their risk of loss or theft. Knowing where these devices are typically stored when not in use is crucial for managing their security. Similarly, peripherals like external hard drives, USB drives, and other portable data storage tools should be accounted for. These small, easily misplaced items can become major vulnerabilities if not properly secured or tracked.

For cloud services or internet-based resources, such as email hosted through Microsoft 365 or data stored on platforms like Google Drive, the physical location of the infrastructure may not be directly under your control. In these cases, designate the location as "cloud." While the physical security of these systems is managed by the vendor, it remains your responsibility to ensure that your provider adheres to appropriate standards of data protection.

Identify any assets accessible remotely

Once you've documented your critical assets and their locations, the next step is to identify which of them are accessible remotely. Remote access introduces unique security risks. Determine which systems can be accessed outside the office and how that access occurs. Access methods might include VPNs, remote desktop software, or web portals.

Your internal network needs layered security similar to an airport. Access should be controlled at entry points, with additional screening inside. Critical systems—like planes on a tarmac—should be tightly restricted. Users should only be able to access the specific systems they need, and only for as long as necessary. Uncontrolled remote access points are like unattended gates, increasing security risks.

Additionally, review the devices used for remote work. Company-managed devices are easier to monitor and secure. If personal devices are permitted, those should be noted on the list, as they will need their own set of guidelines.

Assess infrastructure and building access controls

The next step is assessing infrastructure and building access controls. Begin by verifying that access to your building or office is restricted to authorized personnel. Examine entry points and determine what security measures are in place. For example, do employees need to scan a badge or enter a PIN to access locked areas? Are surveillance cameras monitoring entrances and exits? At some office buildings, a security officer may check IDs or verify visitor credentials.

Next, focus on what access controls protect your critical network infrastructure. Look at the physical location of your primary internet connection and core networking equipment. Is the primary internet line routed through a secure area, or is it easily accessible in a public or lightly monitored space? Servers and other critical devices should never be located in open office environments where they can be accessed freely. Instead, they should be stored in a locked room or cabinet with restricted access to prevent tampering or accidental damage.

Additionally, evaluate the environmental conditions of the network room:

- Is it properly cooled, or is the door propped open with a fan because of overheating?
- For larger server rooms, is airflow optimized with a structured cooling layout, or is cold air simply mixing with hot exhaust and reducing efficiency?
- Beyond temperature controls, are there fire suppression measures in place to protect equipment?

- If the power goes out, is there reliable backup power to keep devices running?
- Does the setup also include a power conditioner to stabilize voltage and prevent fluctuations from shorting out expensive equipment?

Policy Assessment

Check access to sensitive information

Think of your business like a nightclub. You have VIP areas into which only certain people should be allowed. But what if former employees, vendors, or temporary contractors still have their all-access passes? If you wouldn't let a fired bouncer keep a key to the side door, why let old accounts retain access to your data? This doesn't just include employees, either; contractors, vendors, and other third parties might also have been given accounts to complete a project.

Start by creating a comprehensive list of everyone with access to critical systems or sensitive information. This should include current employees, temporary staff, contractors, and vendors. Pay special attention to old accounts, which are often overlooked but may still be active. For instance, a vendor might have been given remote access credentials to set up or maintain a system, yet their account still exists long after the job is done. These dormant accounts represent significant risks, as they are typically not actively monitored but can still provide a potential entry point for attackers.

To streamline this process, your IT team can generate a full inventory of all accounts, their assigned groups, and the last login date. This will ensure no account is missed and help you identify access that may no longer be necessary. The fewer individuals with access to sensitive information, the smaller your attack surface becomes. By limiting access, you reduce the chances of accidental or malicious misuse and make it easier to monitor for unusual activity.

Collect written security policies

Written security policies act as your business's rulebook for protecting sensitive data, managing risks, and ensuring employees follow best practices. But outdated or unclear policies can leave gaps. This section focuses on evaluating the current state of your written security policies, whether they already exist or need to be created from scratch.

Start by collecting all written security policies your business already has. These might be standalone documents or part of an employee handbook, IT instructions, or compliance documents. Common titles might relate to password and authentication policies, acceptable use policies, remote work or bring-your-own-device (BYOD) policies, data classification and handling policies, or incident response plans.

If no formal policies exist, that's okay! Start by documenting any unwritten rules or common practices that employees follow regarding security. Even if you've never officially written down a policy, chances are your team already operates with some assumed security guidelines. These informal rules can reveal inconsistencies, risky habits, or gaps that should be addressed with official policies.

For example, employees may instinctively check with a manager before opening suspicious emails, which is a good security habit to have. However, without a formal policy, not everyone will follow it consistently, and new hires may not even realize it's an expectation. Similarly, if different teams have their own ways of managing access to shared files, that lack of standardized procedures can create vulnerabilities. By documenting these unofficial security practices, you'll get a clearer picture of where policies need to be developed. (If you find that your business doesn't have any structured security policies yet, you can skip the next section on reviewing existing policies.)

Review written security or acceptable use policies

Once you have the written policies in front of you, assess them for the three C's: current, clear, and comprehensive. While those policy documents may have been appropriate when first created, they might no longer align with current systems or practices. For instance, if your organization has recently adopted cloud-based tools or allowed remote work, these changes may not yet be reflected in the policies.

As you review your security policies, use the checklist below. Keep a pen or highlighter nearby to mark any sections that are outdated, unclear, or missing important details. These are the areas you'll want to revisit and update later.

- **Identify any sections that are not current, clear, or comprehensive.** If a policy references software or hardware no longer in use, those sections need to be updated or removed. Similarly, pay close attention to areas that are vague or confusing. A strong security policy should leave no room for interpretation. Employees must clearly understand what is expected of them and how to comply.
- **Evaluate whether any essential topics are missing from your policies.** *Note any policy gaps on the checklist and which policy they belong to.* Here are some examples:
    - Do they address secure remote work practices, like the importance of using a VPN or avoiding public Wi-Fi for work tasks?
    - Is there guidance on creating strong passwords or using multifactor authentication?
    - If your business relies on collaborative platforms like Teams or Slack, do the policies clearly define acceptable use, including guidelines for sharing files and sensitive information?
- **Consider how accessible and user-friendly the policy is.** Documents written in dense, technical language can confuse employees and lead to unintentional noncompliance. Aim for a tone and structure that are clear, concise, and free of unnecessary jargon. The objective is to make the guidelines easy for all employees to understand and follow, regardless of their tech-

nical expertise. *On the checklist, note any policies and sections that could benefit from a language cleanup, in addition to highlighting them on your physical printout.*

- **Finally, involve key stakeholders in the review process.** Your IT team, HR department, and even a few employees who use these systems daily can offer valuable insights into gaps or areas that need clarification. By incorporating their feedback, you can ensure your policies are both practical and comprehensive.

Deep dive into password and authentication practices

Passwords act as the keys to your business's most sensitive information, yet not all keys are created equal. A weak or poorly managed password is like using a rusty old lock on your gate. It might slow down an intruder, but it won't stop them. And using the same password everywhere is like making copies of your house key and scattering them around town. Since your passwords are the keys to your kingdom, it's worth taking a deeper dive into your password policy.

Start by evaluating whether strong passwords or passphrases are required for all devices and accounts. Examine your current password policies to determine the minimum length and complexity required. Are passwords an appropriate length? Do they mandate a combination of upper and lowercase letters, numbers, and symbols? Implementing these requirements helps ensure a baseline level of security across your organization. Additionally, password policies should enforce regular expiration intervals, prompting users to update their credentials periodically. This practice ensures that old, potentially compromised passwords are retired before they can be exploited.

Should You Still Rotate Passwords?

NIST no longer recommends routine password expiration—but that's based on the assumption that your business monitors for breaches, enforces unique passwords, and uses multifactor authentication. Most small businesses aren't there yet. Without periodic changes, an old password leaked in a breach could stay active for years.

Thankfully, password managers make complex, unique passwords easy to manage. Requiring password updates once a year, especially for critical systems, strikes a smart balance. It reduces long-term risk without overburdening your team.

Document your password requirements, keeping in mind that they may vary by platform. For example, your customer relationship management (CRM) software might enforce less strict rules than your email system, while your workstations may have entirely different criteria. These inconsistencies can create security gaps, making it essential to understand and standardize password requirements wherever possible.

In addition to strong passwords, authentication practices should include multifactor authentication wherever possible. Multifactor authentication, or MFA, is like having a smart deadbolt on your front door, plus a standard lock on the handle. Even if someone manages to steal or guess your key (password), they still need a second layer of verification—like a fingerprint scan or a security code—to get inside. Review whether MFA is enabled for all critical accounts you've listed, including email, banking, payroll, and other platforms containing sensitive information.

Flag all accounts that currently do not have MFA enabled. While implementing MFA may require some minor adjustments to workflows, the enhanced security it provides is well worth the effort. Most platforms offer straightforward instructions for setting up MFA, and your IT team can assist with organization-wide implementation to ensure a smooth transition.

Also, check whether employees are saving passwords in web browsers like Chrome or Firefox. While browser-based password storage is convenient, it poses security risks. *Verify whether your business uses a dedicated password manager, or if employees rely on saved passwords in other applications (web browsers, Excel, etc.).*

People Assessment

Review employee training materials

Training programs play a critical role in helping employees recognize and respond to potential threats. During the assessment, review both internally developed materials and third-party awareness programs to ensure your organization is actively equipping staff with the tools they need to support your security efforts. *If you don't currently have any security awareness training materials in place to support these policies, make a note in your checklist and skip the rest of this section.*

Start by examining your current training materials, including presentation slides, handouts, and online modules. Are these materials engaging and accessible to employees at all levels of technical proficiency? Effective training should be clear, free of unnecessary jargon, and grounded in practical, real-world examples. Overly technical or abstract content can reduce the effectiveness of the program and make it harder for employees to absorb and apply what they've learned.

Next, assess the relevance of the information provided. Cyber threats evolve quickly, and infographics that were new two years ago are likely ancient history today. Review the content to ensure it addresses current threats. For example, does the training teach employees how to identify increasingly sophisticated phishing emails, or is it still featuring the Nigerian prince? If anything is over a year old since the last revision, note which document needs to be updated.

The Employee Cybersecurity Training Checklist, provided for download at smcyber.com/satquiz, can help you systematically evaluate your current training program. This checklist includes questions such as: Are all employees required to complete cybersecurity training? How often is the training conducted? Does the training address both digital and physical security practices? By completing this supplemental checklist, you can identify weak points in your program and prioritize areas for improvement.

As you review your training materials, consider whether the content is tailored to different employee roles. For example, a PR specialist managing external communication faces different security risks than a finance manager handling sensitive payment data. Customizing the training to address these specific responsibilities ensures the content is relevant and more likely to be retained by employees. *List any role-specific training materials on the checklist.*

Additionally, check that training materials include actionable steps employees can take when they encounter a potential threat. While recognizing a phishing email is important, employees also need explicit instructions on how to report it to IT and avoid clicking on malicious links. Reinforcing these procedures during training and making them easy to remember and follow turns knowledge into effective action.

Check for unsecured devices or documents

Begin by walking through the office and looking for easily overlooked risks, such as unattended computer screens. Employees who step away from their desks without locking their computers leave an open door for unauthorized access. Similarly, passwords written on sticky notes left in plain sight are significant risks. Consider the printer area: Is it common for employees to forget documents sitting in the tray for extended periods? Inspect desks for sensitive documents that are casually left out instead of being stored properly. These simple oversights can lead to unintentional exposure or malicious exploitation by visitors or unauthorized staff.

As you inspect the workspace, note any devices in vulnerable locations. Are desktop computers freestanding and easy to steal if someone breaks in after hours, or are they secured to desks with locks? Small, portable devices like smartphones, USB drives, and external hard drives are also attractive targets for theft, as they can contain sensitive information.

Mobile devices add another layer of complexity, especially with the rise of remote work and employees using personal devices for business purposes. Ask your IT team for a report on their mobile device management (MDM) policies. Tools like these are essential for enforcing standards like encryption and patching. Are mobile devices centrally managed through MDM software? If a laptop or phone is lost or stolen, does your IT team have the capability to remotely wipe the device to prevent data from falling into the wrong hands?

Going back through your list of critical devices, assign everything a physical risk score from one to three. Note that this differs from the priority tier assigned earlier in the "Identify and inventory critical assets" section. The priority tier ranks how important a device is to operating your business. The physical risk score is the chance of something actually happening to that device. A score of one is a high-risk asset, like the sales manager's laptop or leadership's smartphones. These have a high risk of damage or theft. A score of two is moderate risk, and three is low risk. A spare workstation in dispatch, or maybe even your primary server, could be risk three assets. If the server room is locked with its own climate control, battery backup, and dedicated fire suppression system, there is a low risk of theft or damage to that hardware.

Review backup configurations

Whether dealing with accidental file deletions or full-scale cyber-attacks, a reliable backup system ensures your business can minimize disruptions and recover quickly. Now, you might wonder, *Why are backups included here instead of under "Technology" or "Policies"?*

The answer is because most backup failures happen because of human error, not technical glitches or policy gaps. Issues often stem from poor training, unfamiliarity with backup software, or ignored alerts. While technology and policies provide the tools and structure, people ultimately determine whether those systems are used correctly. If IT employees don't know how to operate or monitor the backup process, the entire system can quietly fail and render your disaster recovery plan worthless.

Start evaluating your backup strategy by verifying the configurations of your backups. Are all critical systems and data being backed up regularly? Confirm that the backup schedule aligns with the needs of your business. Some systems may require daily backups, others might get by with a weekend backup, and some data might need hourly copies of changes. Additionally, ensure that backups include not only files but also system configurations, applications, and databases. Your Core Tier assets should all have frequent backups.

Your broader backup strategy also warrants close attention. Industry standards recommend the 3-2-1 rule: maintain three copies of your data, store them on two different media types, and keep one copy offsite. This approach ensures resilience against a variety of risks, from local hardware failures to large-scale natural disasters.

The location where backups are stored is equally important. Some backups may reside on local office drives, while others might use cloud-based services. Each location presents unique security and accessibility considerations. Local backups can provide faster restoration times but are vulnerable to physical damage or theft, whereas cloud backups offer remote accessibility and disaster recovery capabilities but depend on encryption and fast network speeds. If cloud storage is used, does it retain previous versions of files to prevent corruption? Understanding the storage type of each backup helps you identify gaps and determine if additional solutions are required to enhance your strategy.

Retention policies also play a critical role in ensuring long-term data integrity. Backups should be kept for durations that align with your business's operational and regulatory requirements. Some industries mandate extended retention periods to maintain compliance. Your storage capacity must support these requirements, so evaluate whether your current solutions can handle the necessary retention periods. If they fall short, identify the affected jobs that need additional storage space. Retention policies should balance the need to preserve sufficient historical data with the efficient use of storage resources.

Review restoration procedures

Testing restoration procedures is just as important as creating the backups themselves. A backup that hasn't been tested is like a fire extinguisher you've never checked: It's great to have for peace of mind, but if you wait until the fire starts only to find out it doesn't work, it's already too late.

Start by examining the procedures and runbooks that detail how to restore data, from a single file to an entire server. These documents should provide clear, step-by-step guidance for IT personnel or other responsible parties to follow during the restoration process. In the absence of detailed instructions, valuable time can be lost during an emergency as teams attempt to determine the next steps. *Document any gaps or missing procedures on the checklist.*

Backup software often includes a utility that monitors the integrity of your backups. Instead of fully restoring every file, this tool scans backup sets for signs of corruption and can often repair issues automatically. Verify whether your backup solution includes this functionality and confirm that it's enabled and configured correctly. A corrupted backup is effectively worthless, so regular integrity checks add an extra layer of assurance to your data recovery plan.

Disaster recovery testing is the only way to know your backups will work when it counts. This practice involves periodically restoring files or servers to ensure that everything works as intended. But it's not something you do once and forget about. Systems change, data moves, and what worked six months ago might fail today. *If disaster recovery testing isn't already part of your IT team's routine, note that on the checklist.*

Your IT team should also be able to provide you with estimated recovery times for each backup job. While these times are often based on ideal conditions and should be taken with a grain of salt, they give you a ballpark timeframe in the event of a failure. Understanding how long business operations could be disrupted allows you to create realistic contingency plans to manage downtime.

Wrapping Up

Completing this cybersecurity assessment gives you a clear picture of where your business stands today. It reveals the strengths and the gaps across your systems, policies, and people—insights that are essential for making smart, focused improvements. But identifying issues is only the first step. Real progress comes from acting on what you've found.

As you go through your cybersecurity assessment, you may find certain risks or vulnerabilities that you don't have the time or expertise to fully address. That's okay—most small businesses don't have the resources to handle everything in-house. Identifying security gaps is a great first step, but sometimes closing them effectively requires outside expertise.

Right now, your focus should be on documenting security gaps and prioritizing risks. If you suspect some areas may require external expertise, make a note of them as you complete your assessment. That way, when you're ready to bring in third-party support, you'll already have a roadmap. For guidance on when and how to partner with cybersecurity professionals, see Chapter 12.

Whether you handle changes in-house or bring in outside help, this assessment lays the groundwork for stronger decisions and better outcomes. Now that you know where things stand, it's time to shore up your first line of defense: your physical infrastructure.

Next Steps

- ☐ Lock down one "open door" today: Secure a server closet, router shelf, or unmanned PC before you head home.
- ☐ Ask your IT contact when a server backup was last successfully test-restored.

Resources:

The Security Self-Audit Checklist is a spreadsheet that mirrors this chapter, guiding you through assets, access, backups, and training in one tidy list. Fill it out as you read and identify hidden risks for action in later chapters. Download it from smcyber.com/assessment.

PILLAR ONE

TECHNOLOGY

CHAPTER 5

PHYSICAL SECURITY

Now that you've assessed your assets and their current protections in Chapter 4, this chapter will guide you in selecting and implementing appropriate physical security controls to address your identified gaps.

In cybersecurity, the focus is often on flashy tools and digital defenses. But no matter how advanced those systems are, they all rely on physical infrastructure to function. Servers, firewalls, network cables, and workstations can't protect themselves. If someone can unplug, steal, or tamper with them, many of your digital defenses can be bypassed. Physical security is a foundational part of the Technology pillar of your cybersecurity program, and it's one of the most frequently overlooked. Now that you've assessed your business's risks in the previous chapter, it's time to address those gaps, beginning with the hardware and physical spaces that keep everything running.

Throughout the following chapters, we'll compare your business to a city to help make things easier to understand. In a city, the physical land is the main asset. To preserve this land (and the buildings that sit on it), controls are put in place to keep vehicles and pedestrians in defined areas and funnel them into determined openings. For plots of land where buildings sit, there may be a curb or fence surrounding them to keep vehicles on the street and guide people to the front door. Just as a city needs clear boundaries and controlled access points to function smoothly, your business requires similar protections to ensure security at every level.

If you completed the assessment in the last chapter, you likely identified areas where your physical security could be stronger. Whether it's an unlocked server closet, an unmonitored reception desk, or a lack of security cameras, now is the time to fill the gaps. Let's start by breaking down the most critical physical security risks and how to address them.

Why Physical Security Matters

Physical security encompasses the measures and practices needed to protect physical assets from unauthorized access, damage, or theft. While much of cybersecurity is centered on virtual threats, physical security ensures that the infrastructure supporting your data, devices, and critical operations remains protected. For small businesses, this connection is particularly vital. With fewer locations and limited redundancy, they often depend on a single site to house servers, networking equipment, and essential records. This concentration makes physical security not just an operational issue but a matter of survival.

Now, imagine the phone ringing early one morning with devastating news: A fire has engulfed the office building where your business operates. Without a robust disaster recovery plan and effective physical safeguards, the aftermath could result in the loss of customer data, operational continuity, and the equipment your business depends on. Insurance might cover some of the financial losses, but the downtime and rebuild costs could be devastating.

Similarly, overlooking physical security in third-party vendors can be just as perilous. For example, a failure to vet a cleaning company properly might lead to a rogue employee walking out with your main server. Appropriate physical controls can reduce the risk of an incident caused by a third party.

Physical security gaps often overlap with insider threats and social engineering attacks. Picture this: An individual posing as a job candidate or vendor is given access to your office during an interview. While everyone's attention is elsewhere, they discreetly plug a USB drive into an unmonitored workstation, deploying malware that can steal sensitive information or cripple your network.

For small businesses, deploying physical security measures is a necessity. It acts as the first line of defense against incidents that could threaten your business's operational and financial stability. Implementing measures such as surveillance, restricted access, and disaster recovery plans into your broader cybersecurity strategy protects your organization from both foreseeable risks and the unexpected.

The Role of Physical Security

Physical security is the gatekeeper for all your other cybersecurity defenses. Without strong physical protections, even the most advanced digital safeguards can be easily compromised. The interplay between physical and cyber defenses shows how weaknesses in one area can directly expose vulnerabilities in the other. This relationship is especially vital for small businesses. A breach of physical premises could undermine otherwise robust network defenses.

Physical security is the first line of defense in cybersecurity because it often represents the path of least resistance for an attacker. Why spend days or weeks conducting reconnaissance, exploiting vulnerabilities, and attempting to hack into a well-secured server when the same network can be compromised more easily by exploiting weak physical controls? A determined attacker could bluff their way into an office building during off-hours. Once inside, they could connect a rogue device directly to the internal network, bypassing firewalls, intrusion detection systems, and other virtual defenses by operating within the trusted environment.

This simplicity and efficiency make physical compromise an appealing option for malicious actors, especially in targeted attacks. With physical access, an attacker can deploy tools like keyloggers, rogue wireless access points, or even laptops configured to extract data in a matter of minutes. The speed at which such attacks can occur underscores the importance of implementing robust physical security measures. Layered defenses are essential for deterring unauthorized individuals and identifying suspicious activities before they escalate.

Protecting your physical devices isn't just about preventing theft—it's about controlling who can access your business's critical systems. Strong physical security measures act as barriers, making it harder for unauthorized individuals to get in and giving you more time to detect and respond to threats. Just like with cybersecurity overall, layered defenses work best. The more physical obstacles you put in place, the harder it becomes for an intruder to reach sensitive equipment.

Understanding Physical Security Controls

The selection of physical controls should be tailored to the specific nature of the business, the value of the assets being protected, and the risks involved. In the following sections, we will examine various types of physical controls, each offering unique strengths and applications to meet the diverse needs of organizations.

Fencing, gates, bollards, and lighting form a strong first line of defense for the exterior of a property. While small businesses may not expect elaborate heists, these measures act as deterrents, particularly in areas vulnerable to opportunistic crime. Fencing and gates help control access by restricting entry to authorized points, while bollards provide a physical barrier against vehicular attacks or smash-and-grab incidents, preventing direct access to buildings. Properly positioned lighting enhances visibility, eliminates potential hiding spots, and discourages unauthorized activity after dark. It also supports surveillance systems by improving the clarity of captured footage, making it easier to identify and respond to security incidents.

Locks and keys are some of the simplest and oldest forms of physical security, yet they remain highly effective when used correctly. High-quality locks can secure doors, windows, and cabinets, providing a reliable barrier to sensitive areas. To maximize their effectiveness, businesses should establish key management policies to track who has access and ensure accountability. In the event of a lost or stolen key, replacing locks promptly is essential to prevent a single compromised key from endangering the facility.

Electronic access control systems, like keycards, biometric scanners, and PIN codes, can give you tighter control over who can access specific areas. Unlike traditional locks, these systems can keep track of who entered, when they came in, and how often. That audit trail can be a lifesaver if something goes missing or a breach occurs. Biometric options, like facial recognition or fingerprint readers, take it a step further by using traits that are much harder to fake. They're not foolproof, but they make it much harder for someone to break in using just a cloned keycard or guessed PIN.

Security guards and mantraps add dynamic and active layers to physical security. Security personnel provide human oversight, capable of responding in real time to suspicious activities or attempted breaches. Mantraps, which are small rooms requiring successful authentication to enter or exit, ensure that only authorized individuals can pass through. When used together, these controls offer a highly effective solution for safeguarding high-value areas or sensitive operations.

Security cameras provide continuous surveillance, acting both as a deterrent and as an essential tool for investigating incidents. High-definition systems can monitor wide areas while still capturing enough detail to identify faces and license plates. But footage is only helpful if it's actually being recorded, stored, and reviewed. A surprising number of businesses install cameras and then forget about the storage or never check the footage until it's too late. Make sure your system retains video for at least thirty days, and that someone is trained to retrieve and review footage when needed.

Motion detectors, alarms, and intrusion sensors add an important layer of responsiveness. If someone forces a door or enters a restricted area unexpectedly, these systems can trigger audible alerts and send notifications to designated personnel. They're useful in areas that aren't regularly occupied, such as storage rooms, back offices, or server closets, where unusual movement should always be taken seriously.

Device locks, locking server racks, and file cabinets may seem basic, but they play a critical role in keeping your most important equipment and data out of the wrong hands. If someone gets past your front door or into a secure room, these physical barriers can still slow them down or stop them entirely. They also protect against internal risks, like a disgruntled employee or after-hours tampering. Locking things down isn't just about theft; it's about limiting how much damage someone can do if they're somewhere they shouldn't be.

Requiring visitors to sign in and stay with an employee might seem like overkill, but it's one of the simplest ways to keep tabs on who's in your building. A clear visitor log and escort policy helps prevent someone from wandering into sensitive areas unnoticed. It also adds a layer of accountability: If something goes missing or goes wrong, you'll know exactly who was on-site and when.

Prioritizing Physical Security Controls

With a clear understanding of these physical controls, the next step is to apply this knowledge to create a tailored security posture for your business. This process builds on the assessment checklist you developed earlier, where you prioritized your physical assets based on their importance. By organizing these assets into three distinct categories, you can take a strategic approach to physical security. This ensures that the strongest measures safeguard the most critical systems, while still maintaining practicality and efficiency for less sensitive areas.

To apply layered physical security effectively, you'll need to group your assets based on how critical they are to your business operations. This ensures your strongest protections are focused where they matter most, instead of wasting time or money securing things that don't need it. Think of it like triage in an emergency room: The most critical systems get immediate, high-priority attention, while lower-risk equipment gets lighter safeguards that still make sense for its role. Not everything needs a keycard reader and surveillance camera—but your main servers probably do.

Review your asset list from the Chapter 4 assessment.

The Core Tier should include your most critical assets—those whose compromise could cause severe operational, financial, or reputational harm. These assets require at least three layers of physical controls to ensure their security.

The Operational Tier includes important assets that keep your business running but aren't as sensitive as the most critical ones. These should have at least two physical security measures in place.

Finally, the Support Tier includes the least critical assets, such as general employee workstations or printers. While these systems still require security, a single physical control, such as restricting building access with a keycard system, is usually sufficient.

Adopting a tiered approach to physical security allows you to align protection levels with the criticality of your assets while optimizing the use of resources. This ensures that your most critical systems receive comprehensive protection without overburdening less sensitive areas with excessive measures. The ultimate goal is to create a cohesive, layered defense that protects your business effectively and efficiently.

Building a Comprehensive Physical Security Strategy

When choosing physical security controls, focus on what makes the most sense for your business. Would an electronic access system help you manage who enters sensitive areas? Could motion detectors or alarms be placed strategically to catch unauthorized activity? Or, would a combination of fencing and lighting provide a stronger deterrent against intrusions?

The number of security measures recommended for each category is a starting point, not a limit. These layers form a foundation, but businesses facing higher risks—like those in high-crime areas, handling sensitive data, or operating in regulated industries—should add extra protections where appropriate. Not every asset will need the same controls, but each one should be secured in a way that reflects its importance and exposure. The goal isn't to check a box, but to build a strategy that matches your real-world risks.

For each Core Tier asset, take a moment to evaluate which physical security controls would be the most effective and practical. Your goal is to choose three complementary layers that protect the asset from different angles—such as access restriction, visibility, and tamper resistance. For example, you might use a keypad lock on the server room door, a locking rack, and a camera system to monitor activity. Pay attention to how each control reinforces the others. *Record your chosen controls on the checklist.*

Once you've locked down your most critical assets, shift your focus to the Operational Tier assets in your checklist. These systems may not be business-ending if compromised, but they're still important to day-to-day operations and deserve meaningful protection. Your goal here is to implement at least two layers of physical security for each asset.

When selecting controls, look for opportunities to reuse or extend protections you already have in place. For instance, if keycard access already secures a staff work area, you can add a second layer by locking equipment inside a cabinet or physically securing it with a cable lock. *Note which controls you will use on your Operational Tier assets.*

With protections in place for your critical and moderately critical assets, the final step is securing the remaining Support Tier items on your checklist. While these assets may not be central to daily operations, they still need a basic level of physical security to prevent opportunistic theft or tampering. Just like with the second category, look for ways to build on existing controls while adding safeguards specific to these assets. Many times, an existing control can serve as this layer of protection. This keeps your security approach consistent while simplifying management by using safeguards already in place.

If nothing is protecting these devices yet, start by adding a control that fits the specific risks of each asset. For example, a receptionist's computer in a public-facing area could be secured with a lock cable to prevent theft. These simple but effective measures make it harder for unauthorized individuals to tamper with or steal equipment.

Finally, document all your chosen controls on your checklist for each asset in this category. This final step completes your tiered security strategy, ensuring that every part of your physical infrastructure has an appropriate level of protection.

Wrapping Up

Physical security is the foundation of any cybersecurity strategy. In this chapter, we broke down how layered protections—from fences and locks to cameras—can keep threats out and slow intruders down. It's a simple truth that even the best digital defenses won't matter if someone can walk off with your equipment.

With your physical perimeter in place, it's time to secure what's inside. Next, we'll focus on the devices that keep your business running and show how to protect them against today's digital threats.

Next Steps

- ☐ Spot-check one security camera: Is it recording and positioned effectively?
- ☐ Identify one critical physical asset and list its current protections (e.g., server in locked rack, in locked room).

Resources:

The Essential Physical Security Walk-Through Checklist is a room-by-room cheat sheet covering server rooms, visitor logs, camera placement, and more—perfect for a quick audit with your team. Use it to spot gaps fast and assign fixes in minutes.

Download it here: smcyber.com/physical.

CHAPTER 6

ENDPOINT PROTECTION

Continuing our city analogy from Chapter 5, endpoints are the buildings and structures that make up the town. They come in all shapes and sizes; some are large and important, like city hall or the hospital, while others are smaller but still essential, like homes and businesses. Just like in an actual city, these structures are connected to utilities: electricity, water, roads, and communication lines.

Most buildings are stationary and always connected to the city's infrastructure. Others, like RVs, might move from place to place but still need access to power and water at their destination.

Now, imagine if some of these buildings had no doors, weak locks, or broken windows. They'd be easy targets for criminals. If a homeowner left a spare key under the doormat, anyone could walk in and take whatever they wanted with hardly any effort. The same risks exist for business devices that aren't properly secured.

In the digital world, these buildings represent devices: servers, laptops, tablets, and smartphones. These are collectively referred to as "endpoints." Endpoints are often the first target for cybercriminals because they provide direct access to business data and company networks. If a hacker broke into an employee's laptop, it would be the digital equivalent of breaking into city hall and having access to government records.

And just like RVs that travel outside city walls and connect to unfamiliar infrastructure, mobile endpoints like laptops and phones become even more vulnerable when used on public Wi-Fi or untrusted networks. Because endpoints serve as the front doors to your digital environment, protecting them is a critical part of the Technology pillar in your cybersecurity program.

To secure your business's endpoints against cyber threats, you need layers of protection. Firewalls and antivirus software act as locks and security guards, controlling who and what can enter. Encryption ensures that even if a device is stolen, the data inside remains protected—just like keeping valuables in a safe. Patch management helps keep your digital "buildings" structurally sound by fixing cracks and vulnerabilities before attackers can exploit them. Device management tools allow business owners to enforce security settings across all endpoints, ensuring that every "building" in the "city" follows the same construction codes. Finally, access controls act like security badges and restrict entry only to authorized individuals.

By the end of this chapter, you'll understand how to fortify your endpoints, keeping your business city secure and resilient against threats.

Pizza with a Side of Regret

Sarah ran a small bookkeeping business out of her home, managing financial records for a dozen local clients. One evening, after a long day of work, she decided to pick up a pizza on the way home. She parked her car, locked the doors, and ran inside the pizzeria, leaving her work laptop on the passenger seat. It was only a few minutes, but when she returned, her window was shattered, and her laptop was gone.

Panic set in. That laptop contained banking records, tax filings, and payroll information for her clients. Since she hadn't enabled encryption, all of that sensitive financial data was accessible to whoever now had the device. Worse, she had saved passwords in her browser, meaning the thief could potentially access not just files on the laptop but also her clients' accounting software and business banking portals.

Within days, one of her clients called in distress: Someone had attempted to transfer money from their business account using credentials stored on her laptop. Sarah felt devastated. What was supposed to be a quick dinner stop had turned into a financial and reputational nightmare.

If Sarah had enabled device encryption, the data on her laptop would have remained locked and unreadable, even after the thief stole it.

Additionally, implementing device access controls, like fingerprint biometrics with multifactor authentication, would have ensured that even if the thief had possession of her laptop, they couldn't log in. Without those protections, her business data was left completely exposed.

Layering Endpoint Protection

If securing a business were as simple as installing antivirus software and calling it a day, cybercriminals would be out of work. Unfortunately, that's not the reality. Just as a building uses multiple layers of security—locks, cameras, alarms, and safes—endpoint protection relies on overlapping defenses to keep threats out and limit damage. If one layer is breached, the others still stand, slowing the attacker down or stopping them entirely.

Just as a city relies on layered security (toll booths, fences, safes, law enforcement) to prevent chaos, your digital environment needs multiple lines of defense to stay secure. That means protecting both the devices themselves and the sensitive data they hold. It's not enough to block threats at the perimeter. You need to control who gets in, monitor what happens inside, and make sure that if one system is compromised, it doesn't bring the whole business down.

Antivirus and Endpoint Detection and Response Software

A security system that only reacts after a break-in isn't much of a security system—it's a cleanup crew. Traditional antivirus software works this way: It scans for known threats, detects malicious files, and removes them after they've been found. While this was once the gold standard for endpoint protection, cybercrime has evolved, and so has the technology needed to stop it.

Despite its obsolescence, traditional antivirus software is still everywhere. It often comes pre-installed on big-box store computers or bundled with third-party software, giving businesses the illusion of protection.

These outdated solutions rely on signature-based detection, meaning they can only catch known threats. A business that relies on outdated antivirus software is essentially running security theater—it looks like real protection, but it's doing little more than putting on a show.

Modern threats don't always come as a virus file that antivirus software can detect. Attackers use fileless malware, stolen credentials, and real-time exploitation techniques that bypass traditional antivirus programs entirely. That's where endpoint detection and response (EDR) software comes in. Unlike traditional antivirus software, which primarily reacts to known threats, EDR takes a more proactive approach. It continuously monitors for suspicious activity and is designed to detect and shut down attacks before they can cause serious damage.

Think of an antivirus program as a basic security guard checking employee badges at the front door. It can catch obvious threats—like an intruder wearing a ski mask—but if someone blends in with a forged badge, they can slip past unnoticed. EDR, on the other hand, is like a 24/7 security patrol with a guard dog. It doesn't just scan for known threats; it actively sniffs out suspicious behaviors, such as unusual login attempts, unexpected file encryption, and unauthorized software execution. If something looks off, EDR can isolate the device, stop malicious processes, and prevent attackers from spreading deeper into the network.

Since modern EDR solutions already include antivirus functionality, businesses don't need to run separate antivirus software alongside them. However, EDR is a more advanced tool that requires proper setup, ongoing monitoring, and a trained IT team to manage it effectively. It also comes at a higher cost compared to traditional antivirus software.

For small businesses with limited IT resources, a next-generation antivirus (NGAV) solution may be a more affordable middle ground. NGAV offers better protection than standard antivirus software by using AI-driven threat detection, but without the complexity and overhead of full EDR. While it doesn't offer the deep visibility or real-time response features of true EDR, NGAV can still serve as a strong frontline defense for businesses that need improved protection without the cost and complexity of full security monitoring.

But be cautious when IT service providers claim to include "EDR" in their offerings. In most cases, providers are actually offering NGAV solutions but labeling them as EDR to sound more advanced. While NGAV can still offer strong protection, it's important to verify whether your IT provider is offering true EDR or just rebranding an NGAV product. Many so-called EDR solutions lack the core features that make EDR effective—specifically, visibility into endpoint behavior and the ability for a trained security analyst to detect and respond to threats in real time. Without a dedicated SOC or managed detection and response (MDR) service, EDR is usually just another dashboard no one is watching. To cut through the marketing jargon, here are a few questions you can ask your provider:

- Does this solution continuously monitor endpoint activity and store that data for analysis?
- Can it isolate a compromised device from the network in real time?
- Does it include a centralized console for reviewing alerts, investigating incidents, and responding remotely?
- Is a human analyst or SOC team involved in reviewing or escalating suspicious activity?
- Does the platform support threat hunting or forensic investigation after an incident?

Ultimately, choosing the right endpoint protection comes down to business needs, budget, and IT expertise. Businesses with sufficient IT resources should use EDR for optimal protection, while those without may find NGAV a better option. What's important is that businesses move beyond outdated, signature-based antivirus software.

Host-Based Firewalls

A firewall is like the security gate at the entrance of a business complex: It controls who gets in, who stays out, and what kind of traffic is allowed to pass through. While many businesses rely on network firewalls to protect their entire IT infrastructure, host-based firewalls act as an additional layer of defense at the device level to prevent threats from spreading if they make it inside.

Unlike a network firewall, which filters traffic for the entire business, a host-based firewall is installed directly on an endpoint, such as a laptop, server, or point-of-sale (POS) system. Instead of just securing the perimeter, it ensures that only approved connections are allowed, stopping malware from spreading between internal systems (lateral movement).

A regional fast-food chain learned the hard way that a network firewall alone isn't enough to protect every device. Their corporate office had a network firewall, but that didn't protect them when an unsecured POS system at one of their restaurants became the weak link.

Cybercriminals exploited an unpatched vulnerability in the POS system, and because it was already inside the company's network, the attack completely bypassed the corporate firewall. Without host-based firewalls on the other POS systems, nothing stopped the malware from spreading to every restaurant location.

By the time IT noticed the attack, it was too late. The company had to shut down all digital operations for days, manually processing orders while scrambling to recover. If each POS terminal had used a properly configured host-based firewall, the malware could have been contained, stopping the attack from spreading beyond the initial breach.

In addition to stopping internal threats, a host-based firewall also controls outbound communication, preventing malware from phoning home to attackers or transmitting stolen data. Many cyberattacks rely on infected systems secretly connecting to a command-and-control server for instructions. A properly configured firewall can block these unauthorized outbound connections, cutting off the attacker's ability to exploit the compromised device.

Beyond preventing attacks, host-based firewalls can also enforce application rules, limiting which programs can initiate network connections. This reduces the attack surface by ensuring that only trusted software is allowed to communicate over the network.

Most modern operating systems include built-in host-based firewalls, such as Windows Defender Firewall and macOS Firewall, but these are often disabled by default or misconfigured, leaving businesses exposed.

One of the biggest mistakes businesses make is assuming that once a firewall is installed, they're fully protected. But firewalls require ongoing management to remain effective. Over time, employees install new applications, businesses change cloud services, and networks expand. If firewall rules aren't reviewed and updated, security gaps emerge and leave endpoints vulnerable.

Additionally, host-based firewalls should be configured to work alongside other security measures, not against them. If set up incorrectly, they can block legitimate business applications and lead frustrated employees to disable them completely. That's why it's critical to invest the time and expertise needed to configure firewall rules properly, building security that protects the business without impeding day-to-day operations. A firewall that frustrates users is a firewall that eventually gets turned off.

Device Encryption

If cybersecurity were a bank, encryption would be the vault. It keeps sensitive information locked away so that, even if someone breaks through the front door, they can't get to the valuables. Many small businesses assume encryption is something only large corporations need. But in reality, every business needs encryption, whether for customer data, financial records, employee details, or trade secrets.

Encryption works by transforming readable data into a scrambled format using a mathematical algorithm and a unique cryptographic key. This process makes the information unreadable without the correct *decryption key*—essentially a password that unlocks the original content. Without it, all a hacker can see is a jumble of meaningless characters. Encryption can be applied to files, emails, or entire hard drives, and in most modern systems, it runs automatically in the background without interrupting day-to-day use.

Consider Sarah's stolen laptop. Without encryption on the hard drive, the thief could remove it, plug it into another computer, and instantly access everything: emails, saved passwords, financial records, and customer information. But with full-disk encryption enabled, that thief gets nothing but scrambled nonsense.

Beyond protecting stolen devices, encryption is also critical for securing data in transit. Business data frequently moves between employees, offices, and cloud services. If unencrypted, it can be intercepted by attackers using special tools like packet sniffers. But encrypted data remains unreadable to eavesdroppers, protecting sensitive information even if it's intercepted.

For small businesses, encryption should be applied across devices, files, emails, and cloud storage. Full-disk encryption protects an entire device and ensures data is inaccessible if stolen. Both Windows (BitLocker) and macOS (FileVault) provide built-in full-disk encryption utilities that should be enabled on all business machines.

For more granular control, file and folder encryption protects specific sensitive documents, ensuring that even if someone gains access to a system, confidential files remain secure. Businesses that send sensitive information over email should also enable email encryption, which prevents messages from being read if intercepted. Many modern email providers offer this feature, but businesses should verify that encryption is available and employees know how to use it.

Cloud storage should also be encrypted. While many providers encrypt data by default, businesses should confirm that encryption is applied both *in transit* (while the data is moving across the internet) and *at rest* (when the data is sitting in storage on a server). This ensures that files are protected not just during upload or download but also while they're stored in the cloud. For highly sensitive files, encrypting data *before* uploading it adds an extra safeguard in case the provider's systems are ever breached.

One of the biggest misconceptions about encryption is that it's complicated or requires extra steps for employees. In reality, modern encryption is completely transparent to the user. Once enabled, encryption runs silently in the background without disrupting daily work.

For full-disk encryption, employees log in as usual and enter their password. There's no need to manually enter long encryption keys. The system decrypts data on the fly when accessed by an authorized user. Similarly, email and cloud encryption happen behind the scenes, ensuring security without adding extra steps to business operations.

Encryption is one of the strongest cybersecurity tools, but it's only as secure as how it's used. Weak passwords and poor key management can still expose encrypted data. If encryption is enabled but employees reuse weak passwords or store decryption keys improperly, attackers can still break in.

Additionally, encryption standards change over time. Older cryptographic algorithms, such as WEP and SHA-1, have been cracked and are no longer secure. AES-256 encryption remains unbreakable with today's technology, but businesses should ensure they're using modern standards and updating security settings as needed.

Patch Management

Imagine walking through a city and noticing sagging gates and loose ventilation grates on buildings. Most people might overlook them as wear and tear. But to a seasoned burglar, these are subtle clues that no one's paying close attention. They see a property where the locks might be weak, the alarm system ignored, and the security habits lax—making it a prime target.

The same applies to software vulnerabilities on your devices. Every piece of software, from operating systems to web browsers, contains code that can have security flaws. When developers discover these flaws, they release patches (software updates) to fix them. But if those patches aren't installed, the vulnerabilities remain wide open. Running your business on outdated software is like leaving the front door unlocked in a bad neighborhood. Eventually, someone is going to find it and invite themselves in.

I once worked with a property management company that learned this the hard way. They had their own "Dave" (from Chapter 2) who had clicked "Install Updates Later" one too many times and infected the servers with ransomware. The worst part? The attack was 100 percent preventable. Microsoft had already released a patch for this vulnerability a month earlier. But since the update was postponed, the company left its front door wide open.

Patches apply to everything in your IT environment, but some areas pose higher risk if left outdated.

Servers should always be patched first. They store critical data and run the applications that keep businesses running. Cybercriminals target these systems because of the significant impact they can have. Keeping servers updated is essential, but updates can sometimes require temporary downtime. To manage this effectively, it helps to schedule regular maintenance windows—either weekly or monthly—so that updates can be applied without disrupting staff or business activities.

Operating systems should be next. Whether you're using Windows, macOS, or Linux, attackers constantly exploit OS vulnerabilities. Vendors release updates specifically to fix security flaws, so skipping them is like ignoring a recall on a faulty lock.

Firmware updates are another commonly overlooked risk. Routers, printers, and network switches running old firmware can be hijacked. Hackers use unpatched devices as a stepping stone to launch deeper attacks in your network. Firmware updates for network hardware are often overlooked because they require planned downtime for installation. However, the brief inconvenience of scheduled downtime is a small price to pay compared to the chaos and cost of an unscheduled outage caused by a breach.

Web browsers and productivity software are next in line. Since browsers connect directly to the internet, they're a common attack target. An outdated version of Chrome, Edge, or Firefox could allow hackers to inject malicious code just by serving an infected website. Similarly, productivity tools like Microsoft 365, Adobe Acrobat, and Zoom frequently receive security updates.

Finally, security tools like antivirus and endpoint protection software must stay up to date. Most tools will update themselves but should still be monitored by IT to make sure no device falls behind.

Many businesses fall behind on patching because it feels like a hassle. Updates take time, they require reboots, and sometimes they interrupt workflow. But manual patching leads to delays, and delays lead to vulnerabilities.

The solution? Automate updates.

Modern operating systems and software offer automatic updates, ensuring that patches are applied as soon as they're available. Businesses using IT management platforms can use patch management tools to deploy updates across all company devices without disrupting daily operations. Automated patching removes the risk of human forgetfulness, keeping systems secure without requiring constant oversight.

For businesses that can't enable full automation, establishing a scheduled patching schedule is essential. This ensures updates happen regularly without interfering with business hours. Like locking up at the end of the day, keeping systems up to date should be a regular part of the business's weekly routine.

Device Access Controls

Device access controls bridge the gap between a physical machine and the digital world it connects to. These controls define who can access company systems and what they must do to log in. Without proper safeguards, a stolen laptop or an unattended tablet can become an open door to business data. Many people assume that having a strong password is enough, but attackers with physical access to a device can bypass login passwords in a matter of minutes. Through various techniques, a hacker can reset account passwords, access stored files, and take full control of the device—even if the original user has lost their password.

The first step in securing company devices is requiring strong authentication methods. Every business laptop, desktop, tablet, and smartphone should be protected with a strong password or extended PIN. Passwordless authentication methods, like biometrics, provide an added layer of convenience and security.

To prevent unauthorized access when devices are left unattended, businesses should enforce automatic screen locking after a short period of inactivity, typically a couple of minutes. This simple policy stops opportunistic access when a device is left out in the open. If a company laptop or phone is lost, businesses should have the ability to remotely wipe its data. Both Microsoft Intune and Apple's Find My feature offer remote wipe capabilities to ensure that lost or stolen devices don't become security risks.

Beyond software-based security, hardware restrictions are also necessary. Attackers can bypass login screens by booting a computer from a USB device and accessing the data directly. Businesses should restrict external boot options and set up BIOS passwords to prevent unauthorized changes to a device's security settings.

Work and personal devices should also be kept strictly separate. Employees should not use personal laptops or smartphones for company work unless a mobile device management solution is in place to enforce security policies. Likewise, business-owned devices should be restricted from personal use, as personal apps and activities could introduce malware or unauthorized access.

Last, businesses should track and retire outdated devices properly. Old company laptops, tablets, and smartphones should be securely wiped before disposal or locked down if they are being stored for future use.

Spam Filters and Email Gateways

Every day, businesses receive a flood of emails; some are important client messages, some are routine updates, and others are scams designed to steal information. Sorting through all this traffic manually would be overwhelming, which is why businesses rely on spam filters and email gateways to automate the process and keep out unwanted threats.

Think of spam filters like license plate scanners at a fancy parking garage. When a car pulls up, the system checks its plate. If it recognizes it as belonging to a VIP or trusted client, the gate opens immediately, giving them premium parking near the entrance. If it's a normal employee, they're directed to general parking farther away. But if the system doesn't recognize the car or sees something suspicious, the driver might be redirected to street parking, flagged for additional screening, or even outright denied entry.

This is exactly how spam filters work: They analyze incoming emails for known patterns of spam or phishing attempts, checking things like suspicious sender addresses, harmful links, and unusual keywords. Emails with high spam scores are sent to quarantine or blocked entirely, while trusted messages are delivered as normal.

For stronger protection, email gateways combine several layers of defense to stop threats before they reach your inbox. They scan attachments for hidden malware, check links to see if they lead to fake login pages or malicious websites, and use sandboxing to safely open suspicious messages in a controlled environment. If anything looks dangerous, the message is quarantined automatically, keeping your team from ever seeing or clicking harmful content.

While spam filters and email gateways do an excellent job of blocking common threats, no system is perfect. Some phishing emails are cleverly disguised to bypass filters, which is why businesses should also train employees to recognize suspicious emails. (We'll discuss this more in Chapter 10: Training Employees to be Cyber Aware.)

Wrapping Up

Securing your endpoints is like reinforcing every building inside your digital city. Firewalls, encryption, patching, and access controls all work together to keep intruders out and protect sensitive data where it lives. Crucially, these aren't one-time fixes; they require regular upkeep to stay effective against evolving threats.

If managing all these layers feels overwhelming, you're not alone. Many businesses choose to bring in trusted partners to help design and maintain this resilience. With your individual devices secured, the next logical step is to harden the digital highways that connect them all: your network.

Next Steps

- ☐ Ask your IT contact if automated software patching is in place for your team's devices.
- ☐ Confirm what type of antivirus/endpoint protection (e.g., traditional AV, NGAV, EDR) is on your workstations.

Resources:

The Endpoint Security Scorecard highlights which protections are in place across your devices and where critical gaps remain. Use it as a checklist for your IT team to strengthen your endpoint defenses.

Download it here: smcyber.com/endpoint.

CHAPTER 7

NETWORK HARDENING

Now that we've covered securing individual devices, it's time to focus on the digital highways that connect everything in your business: your network. Every email, file transfer, and cloud login depends on a secure and reliable network. But if your network isn't properly managed, it can become clogged, vulnerable to accidents, or even infiltrated by bad actors. Network hardening is a key part of the Technology pillar in your cybersecurity program, and strengthening your network ensures that the systems you rely on stay protected—not just at the edges, but at every point in between.

Going back to our analogy of a city, there are roads, highways, and utilities that connect the buildings (endpoints). Cars and trucks use the roads to carry cargo from place to place. Utilities provide another type of transportation by moving electricity and water to where it's needed. Similarly, your network is the infrastructure that keeps endpoints connected to one another. The internal network of streets within the city connects everything within city limits. Beyond the borders of the city, the network of streets and highways continues to create a worldwide web of cities.

Just as a city uses traffic laws, bridges, and zoning to control movement and maintain order, your business needs tools like firewalls, access controls, and network segmentation to protect its digital borders. Unfortunately, many small business networks are wide open—poorly segmented, minimally secured, and easy for attackers to exploit. Once inside, cybercriminals can often move freely across systems, install ransomware, or monitor business activity without being noticed.

To build a secure network, a business needs a strong perimeter defense, starting with a firewall. While basic firewalls provided by internet service providers offer minimal protection, next-generation firewalls go further by analyzing traffic in real time, detecting threats, and blocking malicious activity before it reaches your systems. Beyond the firewall, businesses can implement intrusion detection and prevention systems (IDS/IPS) to monitor network traffic for suspicious behavior and stop attacks before they cause damage.

Another key strategy is network segmentation, which divides your systems into separate zones to contain threats and prevent them from spreading unchecked. If one part of the network is compromised, segmentation helps keep the rest of your systems protected.

One often-overlooked vulnerability is the wireless network. Wi-Fi isn't just a convenience—it's a potential entry point. Properly securing Wi-Fi connections with strong encryption and restricted guest access ensures that unauthorized users don't get a free pass into your business network.

Finally, as remote work becomes more common, organizations must ensure that employees connecting from outside the office do so safely. Using virtual private networks encrypts internet traffic, preventing hackers from intercepting sensitive data when an employee works from home, a coffee shop, or a hotel. Encryption also applies to all network communications, ensuring that data remains private whether it's being sent internally or across the internet.

By the end of this chapter, you'll know exactly how to find and fix weak spots in your network before an attacker finds them first.

Firewalls and Next-Generation Firewalls

Your internet gateway functions like a border crossing for your organization's network. It acts as the checkpoint where all internet traffic enters and exits. The magic cable from your internet provider is routed through this point and connected to every device inside your business. Whether employees are checking email or attackers are scanning for weaknesses, all data must pass through this digital checkpoint. And just like a real border crossing, if it's left unguarded, anything can slip through unnoticed.

The first and most critical step in hardening a network is installing a firewall or next-generation firewall (NGFW) device between the internet and your internal systems. Most internet providers install a basic gateway device with built-in firewall features, but these offer only minimal protection. Traditional firewalls typically inspect surface-level details like IP addresses and ports. If web traffic is allowed, they let *all* web traffic through—no questions asked.

What is a Port?

A port is like a numbered lane on a highway that carries one specific kind of traffic, like web browsing or email. Even though it all travels over the same internet connection, each type of traffic uses a different lane to get where it's going.

You might already see the issue here and why more powerful detection capabilities are needed. Next-generation firewalls go beyond this one-size-fits-all traffic filtering. They improve security by using deep packet inspection (DPI): a method that examines the actual content of network traffic, not just where it came from or what port it used. This allows NGFWs to go beyond surface-level filtering and understand what the traffic really is.

One major benefit of DPI is application-aware filtering. NGFWs can recognize specific apps, like Dropbox, Facebook, or Zoom, even if they're using standard web ports. This means you can block or restrict risky or nonessential apps without affecting legitimate web traffic. For example, you might block streaming services like Netflix during work hours, while still allowing access to business-critical platforms.

NGFWs also offer more granular, user-based policies, meaning you can create rules based on employee roles or identities rather than just devices. Some models integrate directly with directory services (like Active Directory), allowing you to tailor access rules for specific departments or job functions. They can also enforce protocol restrictions (like blocking outdated or insecure communication methods), monitor bandwidth usage, and prioritize traffic for essential business functions.

Many NGFWs implement a security subscription service to scan traffic for emerging threats and block access to known malicious destinations. While these subscriptions can be pricey, they may be worth the cost to block a ransomware attack from taking hold.

Intrusion Detection and Protection Systems

An intrusion detection system (IDS) or intrusion prevention system (IPS) complements a firewall by actively monitoring network traffic for suspicious or malicious activity. While a firewall blocks traffic based on a fixed set of rules, IDS and IPS tools look for unusual patterns that might indicate a threat in real time. An IDS focuses on detection and alerting: It can recognize potential attacks, but it's up to your team to decide what to do next. IPS takes it a step further by automatically blocking certain types of malicious traffic as it happens.

Additionally, an IDS can provide forensic evidence by logging detailed information about ongoing network activity, which can be extremely valuable when investigating breaches or understanding how attackers behave. Security teams can customize an IDS, allowing them to focus on specific areas of the network or target particular types of threats.

One challenge of an IDS is the risk of false positives. IDS systems sometimes flag harmless activities as threats. This can cause "alert fatigue" in understaffed IT departments, who may become desensitized to alarms. If ignored or misunderstood, genuine threats could slip through. Because of this, response time is a critical issue. Since the IDS itself doesn't block threats, the burden of remediation falls on the speed and efficiency of human intervention.

On the other hand, an IPS is like a bouncer at a club: It not only spots trouble as it's going down but also kicks out intruders before they can cause damage. It can provide real-time protection without waiting for human input. This automated response reduces the workload on IT teams, who no longer need to manually investigate and block each suspicious activity. This makes it a far more valuable system when dealing with high-speed or automated attacks, including ransomware.

Yet, the IPS is not without its own downsides. Like its little brother, the IDS, it carries the risk of false positives. If an IPS mistakenly blocks legitimate traffic, it can disrupt normal business operations (think a bouncer who turns away a celebrity because he doesn't recognize her). This overzealous blocking can create connectivity issues and lost hours of productivity while IT troubleshoots the problem. Additionally, because it continually analyzes traffic before passing it on, an IPS can cause network performance issues, particularly if it's handling large volumes of data. Misconfiguration is another challenge: Properly configuring an IPS is a complex task, and errors can either leave the network vulnerable or result in overblocking of safe traffic.

Because IDS are good at identifying threats and IPS can block them in real time, many businesses use both as part of their security strategy. Most next-generation firewalls include built-in IDS/IPS features, which are usually enough for smaller businesses. Just be aware that enabling these features can affect performance, so it's worth checking whether your device can handle the extra workload.

Larger organizations with more complex networks often need dedicated IDS/IPS systems, along with a team to monitor the constant flow of alerts. Many times, this monitoring is outsourced to a cybersecurity provider through a service called managed detection and response (MDR). MDR offers 24/7 monitoring, threat hunting, and rapid response from trained experts, providing an extra layer of defense without the need to build a full security team in-house.

Secure Network Devices

Every piece of networking hardware—whether it's a router, switch, firewall, or modem—ships with a default username and password that gives access to its settings. These default credentials are supposed to make initial setup easy, often using something simple like "admin" for the username and "password" for the password. From the device's web-based setup page, anyone can change network settings, disable security features, or reroute traffic. Because these default logins are publicly available online, they're often the first thing a hacker will try when attempting to break into a device.

This is why it's important to change the default password as soon as the device is deployed on the network. Leaving the default credentials in place is like locking your front door but leaving the key under the doormat: Anyone checking the most obvious place will get in.

Some devices require strong passwords to be set on first login, but many do not. In some cases, IT departments leave default passwords in place to simplify access during setup or troubleshooting, but this creates a security risk. A strong, unique password prevents attackers from accessing your network settings and reconfiguring them for their own gain.

But security doesn't stop at passwords. Firmware updates are just as critical. Many businesses skip these updates out of fear that something might break, but leaving devices unpatched turns them into sitting ducks for hackers. The longer a device goes without an update, the more likely it is to have an exploitable weakness.

Another overlooked risk is unnecessary services running in the background. Most network hardware is designed to work in multiple environments, so manufacturers ship devices with extra services enabled by default. These settings might support older networking protocols, remote management tools, or file-sharing services that are never used. Each of these unused services is another potential way for attackers to get in. It's like a school creating drop-off zones not just for cars and buses but for helicopters and submarines, too. The larger the attack surface, the higher the risk. If a feature isn't required for business operations, turn it off!

Network Segmentation

Think about how a city is divided into neighborhoods. Each area has its own infrastructure—roads, power lines, and utilities—all connected to the larger city but still operating within local boundaries. Your business network should work the same way. Network segmentation is about dividing your network into smaller, controlled sections instead of letting every device connect to everything. Without segmentation, an attacker who slips in can roam freely, hitting your most critical systems without resistance. But by putting up virtual walls, you can trap threats in one "neighborhood" before they spread.

Think of your network like a major metropolitan area—take Los Angeles, for example. It's made up of distinct regions like Hollywood, Pasadena, and Montebello, each with its own layout, purpose, and infrastructure. Even though they're all connected and part of the same city, each neighborhood serves different needs and manages its own space. Your business network should follow a similar model. Network segmentation creates purpose-built zones that limit how devices and systems interact. Without it, an attacker who gets in can move freely across everything. But with the right structure in place, you can contain threats to just one "neighborhood" before they spread further.

Most small business networks are wide open, meaning every computer, printer, and server exists in one flat digital space. That's like having no walls inside your office, just one giant room where everyone, including visitors, has access to everything. To limit threats, segmentation creates controlled zones, ensuring that certain devices, departments, or users only have access to what they actually need (see figure 1). The finance team, for example, shouldn't share a network with the warehouse. And customer Wi-Fi? That should be completely cut off from your internal systems; the last thing you want is some random guy in the parking lot poking around your business network.

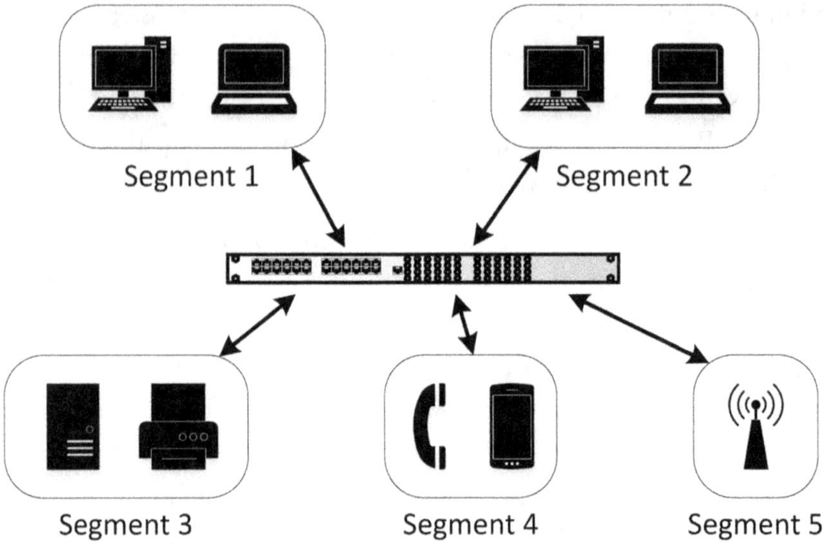

Figure 1: Segmented Network. Each segment is isolated from the others.

One of the biggest benefits of segmentation is attack containment. If ransomware hits an employee's laptop, a segmented network stops it from spreading like wildfire to every other machine. Without segmentation, one click on a bad link could bring down the entire company. That's exactly what happened to a small copier provider: Every department shared the same network, so when a salesman fell for a phishing scam, ransomware spread instantly and locked up every single computer. The entire business ground to a halt. If they'd had network segmentation in place, the infection could have been trapped in the sales department, preventing a full-scale meltdown.

Segmentation also improves network efficiency, similar to how dividing a city into districts improves traffic flow. By isolating local network activity into separate segments, it keeps unnecessary data from flooding the entire system. This reduces congestion and helps ensure that business-critical applications stay fast and responsive, even when other parts of the network are busy.

There are a few ways you can segment your network, depending on your environment. Virtual local area networks (VLANs) are one of the simplest and most cost-effective options. They let you logically divide your network without needing extra hardware. Another approach is to use subnets, which assign different IP address ranges to separate parts of the network. Larger companies with more security needs might use physical segmentation, where switches and routers create physical boundaries between departments or sensitive data. For most small businesses, though, VLANs are the way to go, since they're easy to set up and don't require buying new equipment.

Of course, segmentation isn't a "set it and forget it" deal. It requires planning to make sure systems that need to communicate still can, without opening new security holes. A bad setup could accidentally block access to shared files, printers, or even the internet. But for businesses that want to contain cyber threats and prevent total network meltdowns, segmentation is one of the smartest moves they can make.

Encrypt Communications

When digital devices communicate, they follow specific rules called network protocols, which define how data is sent and received. Think of protocols like different languages spoken within a city: English, Spanish, French, Chinese, etc. While they all allow people to communicate, each language has a unique structure and can't be understood by someone who doesn't speak it. Network protocols work the same way: They define the structure, format, and rules that devices must follow to successfully exchange information.

But not all protocols are created equal. Some are inherently insecure, exposing data to anyone who happens to be listening. When a device communicates using an unsecured protocol, everything is sent in plaintext. This means that if an attacker intercepts the data, they can see *everything*, including sensitive details like usernames, passwords, and confidential business information. This is particularly dangerous on public Wi-Fi networks, where hackers can easily "sniff" or spy on unencrypted data flowing back and forth.

That's where secure network protocols like HTTPS come in. HTTPS encrypts data before it leaves your device, ensuring that it remains private and unreadable to outsiders. This is crucial when transmitting sensitive information like passwords, credit card details, or customer records. If you've ever noticed the lock icon in your browser's address bar, that's your sign that the site is using HTTPS and your connection is secure. If a website doesn't display that lock icon, assume that any information you enter could be visible to anyone on the internet.

While HTTPS is now the industry standard for securing web traffic, some older devices and legacy systems still rely on outdated, insecure protocols. This includes office equipment like copiers, printers, and scanners, which may transmit sensitive data—such as login credentials or scanned documents—without encryption. For example, a multi-function printer sending scan-to-email jobs using unencrypted SMTP could expose usernames, passwords, or document contents to anyone monitoring the network.

Unsecured traffic is like sending a postcard: Anyone who handles it can read the message. Encrypted traffic is more like a sealed envelope: The contents stay private until it reaches the person it's meant for. If your business still relies on older systems that don't support encryption, it's time to update them, or at the very least, isolate them from the main business network.

Implement Virtual Private Networks

Imagine you had your own subway line that ran directly from your house to work. You could walk downstairs, jump into your own private subway car, and walk out into your office a few minutes later. You'd be completely insulated from the outside world, and nobody could look in on you. That's exactly what a virtual private network does for your computer's browsing.

A virtual private network (VPN) is a technology that allows users to securely connect to an internal network over the public internet. It's that private subway tunnel from the outside world straight into your business network. When you connect to a VPN, your internet traffic is encrypted and sent through a secure tunnel to a target device, usually your office's router. From there, the traffic is decrypted and sent on to its destination, as if you were sitting at your desk in the office.

Businesses can set VPN policies so that different users or groups have varied levels of access. Remember network segmentation? Those policies can be applied to VPN traffic as well. For example, a member of the sales team connecting through VPN can only have access to the sales network. These abilities make VPNs an essential requirement for any employee working remotely or needing to access internal resources from outside the office. Implementing a VPN can also help organizations comply with regulations by ensuring data is encrypted and secured in transit.

It's worth noting that many public Wi-Fi networks, particularly in hotels and airports, actively block VPN traffic as part of their network policies. This can be frustrating for users and may cause long troubleshooting calls to the IT helpdesk. The best workaround is to use a mobile hotspot for internet connectivity while traveling and to avoid public Wi-Fi networks completely.

Additionally, VPNs can sometimes reduce connection speeds because of the encryption process. This performance reduction can be noticeable, especially with slower connections like mobile hotspots. Certain VPN configurations or software versions may contain bugs that affect speed. Therefore, IT must carefully configure the VPN and use the correct software versions. For enhanced security, MFA should always be enabled on VPN connections.

Tighten Wi-Fi Networks

Wi-Fi is an often overlooked attack surface in many businesses, even though it's one of the easiest entry points for hackers. Unlike wired connections, which require physical access, Wi-Fi is designed to be broadcast and accessible to anyone within a certain range. Depending on how it was set up, often the Wi-Fi signal will extend beyond the walls of a building and into the street or neighboring buildings. With a homemade directional antenna built from around $25 in parts, an attacker could intercept your signal from up to a mile away. This combination of convenience and extended proximity is what makes Wi-Fi so vulnerable.

Wi-Fi security relies on encryption protocols, and some are safer than others. Never use WEP because it can be hacked in seconds. WPA is slightly better but still not secure. WPA2 is the current standard and is safe for most businesses, while WPA3 is even stronger but not yet widely adopted. If your devices support WPA3, use it. Otherwise, WPA2 with a strong password is your best bet.

Regardless of the Wi-Fi protocol you use, a strong password or pre-shared key (PSK) is essential. And please, for the love of cybersecurity, do *not* use anything related to your business name, location, phone number, or industry! Hackers build custom wordlists using publicly available details, not just about your company but also about your entire field. I once worked with a company that used its main office phone number as the Wi-Fi password—something listed right on their website. A dental office using a password like "Smile2024" or "ToothCare1!" might think it's unique—but attackers are already trying dental-related terms. The same goes for healthcare offices using "Wellness123" or law firms using "LawOffice559." These are the first things a targeted attacker will test, not the last. With basic tools, an attacker can generate a wordlist of thousands of password combinations in minutes—mixing common industry terms, business names, and number patterns like addresses or years. Then they automate the attack, testing each one until something works.

Many businesses choose simple Wi-Fi passwords so clients or customers can connect easily, but that convenience should never come at the cost of security. A better approach is to create a separate guest Wi-Fi network specifically for visitors. Guest networks are isolated from your internal systems and can prevent outsiders from accessing sensitive resources like servers, printers, or workstations. Even employee personal devices that only need internet access (like smartphones) should connect to the guest network, not the main business Wi-Fi. You can also enable content filtering and bandwidth limits on guest access to reduce risk and prevent abuse.

One popular technique to secure a Wi-Fi network is to hide the broadcast name. This is known as security through obscurity, where the network is "hidden" and doesn't appear in the list of available Wi-Fi connections. But this is nothing more than a false sense of security. While hiding the name might prevent casual users from seeing the network, any attacker will easily find it with basic scanning tools. A growing number of devices now even identify hidden Wi-Fi networks as "Unknown Network" in the list of connections. Hiding a broadcast name can add a small component of protection, but it should never be relied upon as a primary security measure. Instead, focus on using WPA2 or WPA3 encryption, strong passwords, and network segmentation to truly protect your business's Wi-Fi.

Email Security

As you learned in Chapter 6, setting up an email gateway is a good start. However, because email is so vital for business communication, it requires extra security. Attackers often try to spoof messages, impersonate senders, or change emails while they're being delivered, so strong security is essential. Three key features—SPF, DKIM, and DMARC—work together behind the scenes to prevent this. Most people outside of IT don't even know these settings exist, but they're crucial for ensuring your emails arrive safely and aren't faked. While your IT team will handle the setup and maintenance, it's important that you understand how they protect your business.

SPF - Preventing Spoofed Messages

Sender Policy Framework (SPF) helps prevent attackers from faking your email domain to send phishing emails that look like they came from your company. Your domain is the portion of your email address after the @ symbol (e.g., in someuser@gmail.com, the domain is gmail.com).

SPF works by publishing a public record that lists which mail servers are authorized to send emails from your domain. When an email arrives, the recipient's mail server checks this record to verify whether the message came from an approved source. If it doesn't match, the email is flagged as suspicious and may be sent to the spam folder or blocked entirely.

This doesn't apply solely to emails sent by employees. Vendors that send emails on your behalf must be authorized, too. Many cloud platforms, like QuickBooks, send emails that appear to come from your domain (e.g., accounting@yourcompany.com sends invoices). To prevent these emails from being flagged as spam, IT needs to add QuickBooks as an allowed sender in your SPF record. Vendors will provide the necessary SPF details, and your IT team should know how to update them.

DKIM - Ensuring Emails Aren't Tampered With

SPF prevents senders from being spoofed, but it doesn't stop an email from being modified after it's sent. That's where DomainKeys Identified Mail (DKIM) comes in.

DKIM protects emails by digitally signing outgoing messages with a cryptographic key. When an email is sent, the mail server generates a unique signature and embeds it in the message header. The recipient's mail server checks this signature against a key stored online by your domain. If they match, the email is confirmed as authentic and unaltered.

Think of it like sending a royal decree carried by a trusted messenger, the scroll sealed with wax by the king's signet ring. Upon arrival, the recipient will inspect the seal. If unbroken, he'd trust the decree's authenticity. But if the seal were cracked or missing, he'd know the message might have been tampered with on its journey.

DKIM works the same way: It prevents attackers from modifying an email's content after it's sent and provides a way for the recipient to verify its authenticity. Without DKIM, a hacker could alter an invoice or redirect a payment without the recipient knowing the email was forged. To keep DKIM effective, businesses should rotate their DKIM keys regularly, just like changing locks periodically.

DMARC - Enforcing Email Security and Reporting Issues

Domain-based Message Authentication, Reporting, and Conformance (DMARC) acts as the third layer of email security, tying SPF and DKIM together into one policy. It tells recipient mail servers what to do if an email fails SPF or DKIM checks.

Without DMARC, it's up to the recipient's spam filter to decide whether to block, flag, or deliver an email that doesn't pass authentication. DMARC removes this guesswork by letting your IT team define how failed messages should be handled—whether they should be rejected, quarantined, or allowed through.

Another key benefit of DMARC is reporting. It generates detailed logs that show which mail servers are sending messages on behalf of your domain. This helps identify unauthorized senders or misconfigurations so IT can fine-tune security settings and spot potential email abuse before it becomes a problem.

To recap, SPF verifies which mail servers are allowed to send emails from your domain, DKIM ensures emails haven't been altered during transit, and DMARC enforces rules on failed messages and provides reports for ongoing monitoring.

If your domain doesn't have SPF, DKIM, and DMARC set up, attackers can impersonate your company and send phishing emails to your customers and partners. Many email providers now block messages that lack these protections, so having them in place is no longer optional.

To check if your domain has SPF, DKIM, and DMARC configured, ask your IT team or use an online tool like MxToolbox. If they're missing, work with IT to set them up. This level of email protection is a must-have to prevent email fraud and protect your company's reputation.

Wrapping Up

Your network is the backbone of your digital operations, and securing it is nonnegotiable. From firewalls and intrusion detection to segmentation and encryption, each layer helps block threats and limit their reach. Email protections like SPF, DKIM, and DMARC are now baseline requirements for preventing impersonation and phishing attacks.

Addressing gaps in these defenses now will help protect your business's systems, finances, and reputation down the road. With your network secured, the next step is controlling who has access to what, because even trusted users need boundaries.

Next Steps

- ☐ Check if your business uses a separate, isolated Wi-Fi network for guests or general employee use.
- ☐ Ask your IT team/provider if your main office firewall is a next-generation firewall (NGFW) and if it has an active security subscription.

Resources:

The Network Security Red Flag Checklist helps you spot common weak points in your business network. Use it to spark the right conversations with IT and prioritize what needs fixing first.

Find it here: smcyber.com/network.

PILLAR TWO

POLICIES

CHAPTER 8

DATA ACCESS CONTROLS

A small business and a high school aren't that different from one another. Imagine a high school where no one has assigned classrooms, the doors to the teacher's lounge and principal's office are wide open, and students can wander wherever they want. Teachers find kids rummaging through file cabinets, janitors get pulled into answering phones in the main office, and a group of unsupervised freshmen just discovered the chemistry lab's supply closet.

It wouldn't take long for things to go terribly wrong.

While this might sound ridiculous for a school, it's exactly how many small businesses operate when it comes to digital access. Nobody has taken the time to decide who should have access to what. Just like a well-run school relies on locked doors, hall passes, and assigned access to resources, businesses need structured data access controls to prevent chaos. Even the best cybersecurity technology won't help if access is wide open, and even well-meaning employees can make mistakes.

In this chapter, we'll break down how small businesses can rein in access control without overcomplicating everything. You'll learn how to limit access to only what's necessary, how to manage employee permissions efficiently, and how to prevent security gaps that hackers love to exploit. Access control is a core part of the Policies pillar in your cybersecurity program. The best technology can't protect your data if the rules around who can access it are too loose or unclear. While this chapter touches on some technical controls, we're building an access control framework at its core: a clear policy that defines who gets access to what, and under what conditions. The technology simply enforces the decisions we define here.

The first rule of access control is simple: Only give people access to what they need.

Principle of Least Privilege

At a high school, the janitor doesn't have a key to the HR file cabinet. The chemistry teacher doesn't have access to student medical records. The football coach doesn't get to review the school's budget. Everyone has access to exactly what they need to do their job, and nothing more.

This same principle applies to cybersecurity.

The principle of least privilege (PoLP) is the idea that employees should only have the minimum access necessary to perform their jobs. Nothing extra. Nothing "just in case." The more access someone has, the more damage can be done by accident, negligence, or outright malicious intent.

Most businesses don't think about access control until something goes wrong. It's easier to leave everything open than to figure out appropriate restrictions. But here's the problem: Unrestricted access can turn small mistakes into major disasters.

Take Danielle, a receptionist at a small farm distributor. She answered phones, processed invoices, and scheduled customer pickups. Since it was a small, family-run office, no one thought much about who had access to what. Employees shared the same logins and a single network folder, and trust was the only security policy in place.

One morning, Danielle opened an email that looked like it came from a supplier. Without thinking twice, she clicked the link. Within minutes, the company's entire server was compromised.

Because there were no access restrictions, the hackers didn't just see the delivery schedule. They could access payroll files, vendor financial details, and the company's entire customer database. What should have been a minor security incident turned into tens of thousands of dollars in financial damages.

If Danielle's account had only been limited to what she needed—scheduling and invoicing—the damage would have been contained. Instead, her over-permissioned account gave hackers the digital equivalent of a master key to everything.

The best way to prevent a situation like Danielle's is simple: Stop handing out master keys. Instead of giving employees unrestricted access, think about what they actually need to do their jobs. A receptionist shouldn't be able to view payroll records. A technician in the warehouse doesn't need permissions to approve vendor contracts. If an employee's job doesn't require a certain type of data, they shouldn't have access to it, period.

When businesses are small, it's easy to manage who has access to what. Manually setting permissions for each employee sounds easy—until you have to do it for dozens, or even hundreds, of employees. That's why businesses need a structured approach. Enter role-based access control (RBAC): a framework that makes permission management simple, scalable, and secure.

Role-Based Access Control

RBAC solves the scaling problem by structuring access around roles instead of individuals. Instead of manually assigning permissions one employee at a time, you define job-based access levels, ensuring that employees automatically receive the right permissions when they start, switch roles, or leave. This approach saves time, reduces human error, and makes access controls scalable. No more juggling individual permission settings or discovering months later that a former employee still has access to critical systems.

RBAC also simplifies security audits and compliance. If a business needs to verify who can access payroll records, RBAC makes it easy to confirm that only the finance and HR departments have that permission.

Without it, access control turns into a mess of one-off exceptions, where no one really knows who has access to what.

Take Brian, a site manager at a small apartment management company. When he was first hired, he only had access to his assigned property's records. But over the years, as he covered for coworkers, helped with leasing, and assisted the finance department with reports, IT kept giving him new permissions without removing the old ones.

By the time he left, Brian had way more access than he should ever have had, including tenant financials, company banking details, security camera feeds, and even payroll records. And because no one was keeping track, his login remained active after he quit.

Then came the breach.

Someone using Brian's old credentials accessed company files. IT scrambled to determine what had been exposed, but the patchwork of accumulated permissions made it nearly impossible to assess the full damage.

Most businesses don't actively monitor permissions, and over time, employees accumulate way more access than they should. This happens all the time: Employees switch roles, take on short-term projects, or change departments, and their old permissions never get cleaned up. This accumulation of unnecessary access is called "permissions creep." Auditing user access at least quarterly helps keep permissions aligned with job roles and prevents unnecessary security risks.

This example highlights the value of RBAC. With a role-based system, as soon as Brian switched departments or took on new tasks, his permissions would have been automatically aligned with his current role. And when he left the company, his access would have been revoked as part of the offboarding process. Instead of relying on manual updates, RBAC would have kept his permissions tied to his job function, preventing permissions creep and mitigating the risk of a breach.

Implementing RBAC

The best way to start implementing RBAC is to map out key roles within your company. Instead of focusing on individuals, look at job functions. Ask yourself, *What systems and data do employees need to do their jobs?* A customer service rep might need access to order histories and client emails, but they don't need payroll records. A warehouse worker might need inventory systems but not customer databases.

Once roles are clearly defined, permissions should be assigned to those roles instead of individual employees. This way, when someone is hired, changes roles, or leaves, their access is automatically adjusted, without constant manual updates.

One of the biggest security pitfalls businesses fall into is granting temporary access without a clear plan to remove it. A manager might say, "Just give them access for now," intending to revoke it later, but in reality, that rarely happens. Temporary workers often receive access for a single project, only for IT to forget about it after they're gone. Over time, these overlooked permissions pile up.

RBAC systems like Active Directory don't natively support automatic expiration for role assignments, so businesses need to be proactive about managing temporary access. Rather than granting ad hoc exceptions, it's better to create designated temporary roles with clearly defined and limited permissions. These roles can then be reviewed periodically and removed once they're no longer needed. For example, IT could create a "Database Project" group that grants a programmer the specific access needed during development. As part of the project's offboarding checklist, they can include a step to remove the programmer from the group or delete the group entirely once the work is complete.

Additionally, businesses should schedule recurring permissions audits to identify lingering access that should have been revoked. For organizations that require more granular access controls, we'll explore additional options later in this chapter.

RBAC isn't about restricting employees; it's about keeping security manageable as your business grows. Without it, access control quickly turns into a chaotic mess of outdated permissions, security gaps, and compliance headaches. But while RBAC organizes access at the job role level, sometimes more precision is needed.

Access Control Lists

Think of RBAC as a ticket to a private event. But just because someone has access to the venue doesn't mean they can go everywhere. Your role—whether you're a guest, event staff, or a performer—determines which areas you can enter.

A general admission guest might have access to the main floor but not the backstage area. And just because someone has a backstage pass doesn't mean they can raid the catering table or mess with the sound equipment. Even within VIP access, there are layers of permissions.

That's where access control lists (ACLs) come in. While RBAC defines *who* should have access based on their role, ACLs define *how much* access someone has within the areas that are granted to them. An ACL is a list of specific permissions tied to a file, folder, or system resource that defines who can do what. It controls exactly who can interact with that item and what actions they're allowed to take, such as viewing, editing, or deleting. While RBAC determines *who* should have access based on their role, ACLs define *how much* access they have. So, while RBAC helps assign the right level of access based on roles, ACLs fine-tune this access, allowing businesses to control what specific actions employees can take with the resources they're permitted to interact with.

At Danielle's farm distribution company, ACLs could have ensured that she only had read/write access to the scheduling folder and read-only access to the customer list. The financial and vendor records would have been completely off-limits to her user account.

Implementing ACLs

Implementing ACLs doesn't have to mean creating complicated structures that slow down productivity. The key is to start simple. Begin by identifying your most sensitive data, such as payroll files, customer financials, and vendor contracts, and ensure only the right people have access.

Once you've identified this critical data, set up ACLs to control who can access what. Just because someone can view a file doesn't mean they should be able to edit or delete it. Sensitive files should be tightly controlled, and they shouldn't be left open for everyone to modify. A well-structured ACL ensures that putting a document in a shared folder doesn't automatically grant everyone permission to change or delete it.

To make things easier, ACLs work best when paired with RBAC. Rather than manually assigning individual employees to every folder's ACL, you can use RBAC groups to simplify permissions. This way, you're aligning permissions with roles, making things much more efficient.

For example, instead of adding each accountant to the accounting folder's ACL, you can assign them to an "Accounting Team" group. The ACL then grants that group read/write access to the files in that folder. A subfolder containing department policies and procedures might grant the same group read-only access, ensuring that only members of an "Accounting Manager" group can change procedure documentation. When a new accountant is hired, IT simply adds them to the Accounting Team group. If they leave or switch roles, removing them from the group instantly revokes their access across all relevant files.

By combining ACLs with RBAC, you can create a system that's not only easier to manage but also much harder for employees (and attackers!) to exploit. You can define access based on roles, then fine-tune what those roles can actually do. At this point, you're probably starting to see just how practical—and powerful—this combination can be.

ACL Risks

While ACLs provide granular control, they also introduce risks if not properly managed. One common mistake is misconfigured storage, where sensitive files or entire databases are left accessible because of incorrect ACL settings.

A striking example of this occurred in August 2024, when Confidant Health, a U.S.-based mental healthcare provider, inadvertently exposed a massive 5.3 terabytes of highly sensitive patient data in an unsecured cloud storage database. The records included therapy session details, medical histories, video and audio recordings, and psychiatry intake notes—all openly accessible to anyone with an internet connection. Without proper ACL restrictions in place, no authentication was required to access this sensitive information. The breach was only discovered when a security researcher stumbled upon the exposed database and reported it.

This kind of exposure is far too common in cloud storage environments. Cloud-based file-sharing platforms like Google Drive, Dropbox, OneDrive, and AWS S3 buckets can default to overly permissive sharing settings, making it easy for businesses to accidentally leave files accessible to anyone. A single misconfiguration, such as leaving a database on "public" sharing instead of restricting access, can lead to catastrophic data leaks, like what happened with Confidant Health.

Another major risk is "permission sprawl," where ACLs become overly complex over time. If ACL settings aren't regularly reviewed and cleaned up, businesses end up with a tangled web of outdated permissions. When ACL structures become difficult to track, even IT staff may lose sight of who has access to what, making it nearly impossible to enforce security properly.

When properly implemented, an ACL works like a well-managed VIP list: It allows only the right people to enter while keeping others out. Once ACLs are in place, businesses should go one step further by enforcing policies that dictate how employees should use their access responsibly. That's where acceptable use policies come into play.

Acceptable Use Policies

At a high school, every student knows the computer lab rules: no gaming during class, no messing with the system settings, and definitely no looking up things you wouldn't want your teacher to see. These rules prevent disruptions, protect the school's network, and ensure that students use the technology responsibly.

Businesses need the same structure for employee access to company systems. That's the responsibility of acceptable use policies (AUPs). An AUP defines how employees are allowed to use company technology, networks, and data, ensuring that security risks are minimized while productivity remains high.

Without an AUP, employees must guess what is acceptable, creating serious security risks. From browsing questionable websites on company devices to using personal email accounts to handle work-related documents, a lack of clear rules opens the door to many insecure practices.

Think of an AUP as the rulebook for how employees should handle company networks, devices, and data, preventing the office from becoming a digital Wild West. It lays out clear dos and don'ts to keep business operations running smoothly while minimizing security risks.

For starters, an AUP sets boundaries on internet use so employees aren't using company resources to visit shady websites, gamble online, or binge-watch Netflix on the clock. It also spells out email and communication rules, making sure sensitive business transactions stay out of personal email accounts, which lack the security protections of corporate systems.

Beyond web and email use, an AUP also covers software installation policies, because letting employees download random apps is a great way to invite malware into the network. It reinforces best practices for remote work, like requiring employees to use a company-approved VPN and avoid unsecured public Wi-Fi while traveling.

Most importantly, an AUP sets the ground rules for handling sensitive data. Employees need to know exactly what's expected when dealing with customer information, financial records, or internal documents. When people have clear guidelines, they're less likely to make security mistakes that could expose the business to unnecessary risks.

Unfortunately, not every business takes these precautions. When policies are unclear or ignored, mistakes happen. That's exactly what happened to a small law firm in 2022, when an employee casually checked their personal email on a company laptop. What seemed like a harmless habit ended in disaster.

The employee downloaded an attachment from an unknown sender. That attachment contained malware that spread through the firm's network, encrypting case files, client contracts, and even financial records.

Had the firm implemented a clear acceptable use policy, restricting the use of personal email on company devices and warning employees against opening unknown attachments, the entire attack could have been prevented. Instead, the business lost critical legal documents, faced costly downtime, and had to pay a ransom to regain access to their files.

One of the biggest challenges small businesses face is getting employees to follow the rules. An AUP only works if employees understand why it matters. Instead of just handing out a policy document, businesses should incorporate AUP training into onboarding, refresher courses, and real-world security awareness exercises.

To make enforcement practical, businesses should combine the Policies pillar with the Technology pillar and use technical controls to back up their policies. If streaming sites aren't allowed on work computers, a web filter should be in place to block access. If unauthorized USB drives pose a security risk, IT should disable USB ports for employees who don't need them. A strong AUP is proactive, not reactive. It stops bad habits before they lead to security breaches.

But sometimes, creating rules isn't enough. Organizations also need a method to enforce them dynamically. That's where attribute-based access control comes in, allowing security policies to adapt based on context, location, and risk.

Attribute-Based Access Control

After a chaotic first week at the hypothetical high school, leadership has realized they need more nuanced rules for accessing certain areas. Now, the janitor can unlock classrooms, but only after school hours. A student can enter the computer lab, but only when a teacher is present.

This dynamic approach mirrors how attribute-based access control (ABAC) functions. Instead of granting access based only on a person's job role (like "student") or group membership (such as the "Chemistry Student" group), ABAC introduces additional rules called *attributes*. In our high school analogy, a student must not only belong to the right group but also meet other conditions: They must be actively enrolled in the chemistry course, supervised by a licensed instructor, certified to handle the equipment, and physically present in the lab during open hours before access is granted. Similarly, in a business setting, ABAC allows you to define rules beyond static job roles. It incorporates attributes like time of day, location, device type, and security status to determine whether access should be granted or denied.

This approach to access control provides much more flexibility than traditional models like RBAC. By defining specific conditions for access, ABAC allows businesses to create more granular and dynamic access rules that adapt to the context of each access request. For example, instead of allowing employees to log in from anywhere at any time, a business can require multifactor authentication whenever a user attempts to access company systems from outside the office network.

With ABAC, access control becomes smarter. Instead of blindly allowing access just because someone is an "accountant," ABAC evaluates multiple factors before granting permission. The system might allow an accountant to access financial records only if they are using a company-issued laptop, connected to the corporate VPN, and working between 8 a.m. and 6 p.m. If they attempt to log in from an unknown device in another country, access is automatically denied—even if everything else checks out.

How ABAC Differs from RBAC

Unlike RBAC, which assigns access based on static roles, ABAC is context-aware, meaning it can adapt security decisions in real time. In July 2024, a financial institution successfully stopped an attempted breach thanks to ABAC policies. The company had configured conditional access rules in Microsoft 365 to block logins from unauthorized locations. One day, the security operations center received alerts of login attempts from Egypt, Pakistan, and Germany, all within ninety seconds. Because ABAC policies were in place, these login attempts were automatically blocked, and the incident was contained within minutes with no compromise of data.

Had the company relied solely on RBAC, these login attempts might have succeeded, since the hackers had stolen valid credentials. ABAC added an extra layer of intelligence, blocking unauthorized access before any damage was done.

Another advantage of ABAC is its ability to enforce temporary access permissions. For example, an IT contractor working on a short-term project can be granted access to specific systems only for the duration of their contract, with permissions automatically expiring when the work is done.

Many small businesses assume ABAC is too complex or costly to implement, but modern cloud platforms like Microsoft 365 already include built-in ABAC features that make it easier to enforce dynamic access controls. Instead of manually setting up every condition, businesses can start with a few key security policies that automatically adjust based on risk factors.

Instead of overwhelming IT staff with complex configurations, businesses should focus on defining practical access rules that align with real-world security risks. ABAC complements RBAC and ACLs by adding an extra layer of intelligence that ensures employees can access only what they need, when they need it, and under the right conditions.

Wrapping Up

Strong data access controls are essential for keeping your systems secure and organized. Open access may feel convenient, but it creates unnecessary risk. Shifting to intentional access—where people only get what they truly need—shrinks your attack surface and strengthens your defenses.

Tools like the principle of least privilege, RBAC, ACLs, and ABAC give you flexible ways to manage access based on roles, rules, and context. Paired with clear acceptable use policies, these controls align technical enforcement with everyday behavior.

Together, they don't just block threats—they help build a more efficient and resilient workplace. Next, we'll look at how to prepare for the inevitable: what to do when something does go wrong, and how to recover fast.

Next Steps

- ☐ Ask your IT team how access permissions are typically assigned (e.g., per person, by job role/group).
- ☐ Locate your company's acceptable use policy (AUP) for technology.

Resources:

This Sample Acceptable Use Policy spells out clear dos and don'ts for internet use, email, software installs, and data handling. Just add your company details and share it with staff. A bonus password policy template is included to help you enforce strong credentials across the board.

Download it here: smcyber.com/policies.

CHAPTER 9

INCIDENT RESPONSE AND DISASTER RECOVERY

Business was humming along as usual for a regional financial agency and its thirty-five staff members one Tuesday. It was after lunch, and everyone was just getting back to work when, suddenly, all the shared drives disappeared and remote desktop sessions froze. Moments later, programs crashed with "Database Timeout" errors, and the internet went down. All work stopped immediately while the thirty-five employees stood around and waited for IT to investigate.

The company's only server had powered off and refused to boot up again. The IT technician discovered that a small but critical cooling component had come loose inside. Without it, one of the server's chips had overheated and become permanently damaged. With no backup hardware available and a two-week wait for a replacement server, employees were sent home while IT scrambled to bring the company back online.

In the 2015 movie *Jurassic World*, the park's catastrophe is described as "an eventuality." The same holds true for your business. Whether it's hardware failure, a cyberattack, or a natural disaster, it's not a matter of *if* something will go wrong, but *when*.

That's why you need structured, documented policies that outline how your business will react under pressure. These form a key part of the Policies pillar in your cybersecurity program, ensuring your team has a clear plan when it matters most. In this chapter, you'll learn how to prepare for disruptions, respond effectively, and recover faster, so your business can weather the unexpected and keep moving forward.

Your Cybersecurity Fire Drill

Incident response is your business's emergency plan for cybersecurity. When a hacker breaks in, a virus spreads, or a phishing email arrives, your team needs a clear, immediate course of action. Just like a fire drill ensures people know where to go in an emergency, an incident response plan will help a business contain damage and restore operations quickly, minimizing the impact of the incident.

Disaster recovery takes this a step further, focusing on how to restore critical systems after major disruptions like cyberattacks, power outages, and hardware failures. Just as schools prepare for hurricanes or fires, businesses must plan how to keep running if servers go down or data is lost. This includes having reliable backups, ensuring rapid system recovery, and developing contingency plans to minimize downtime and keep employees productive.

Incident response and disaster recovery are just two parts of a larger strategy: business continuity planning. A business continuity plan, or BCP, ensures that the entire business can continue operating during and after any disruption, addressing more than just technology. While disaster recovery focuses on restoring systems and data, BCP goes further by keeping customer service, operations, and communication functional even if your main office is inaccessible.

Benefits of Incident Response Planning

One of the major benefits of having a solid incident response plan is minimizing downtime. Time is money, especially for smaller businesses operating on tighter margins. Every minute spent offline or recovering from an incident is a minute of lost productivity, frustrated customers, and delayed operations. Having a well-documented process helps teams quickly identify the issue, isolate affected systems, and begin recovery without scrambling to figure out what to do next.

Incident response also saves costs. Cybersecurity incidents can rack up expenses in multiple ways: emergency IT support, third-party fees, and even ransom payments in the worst cases. Having a preestablished plan can minimize the need for expensive, last-minute solutions. It's far cheaper to prevent escalation than to repair the damage after it's done.

Beyond immediate costs, there's the issue of protecting sensitive data. A data breach of customer information leads to long-term consequences. For businesses subject to regulatory compliance, a breach may result in hefty fines. These penalties often dwarf the cost of developing an incident response plan and can threaten the long-term survival of a business.

Lastly, incident response supports the invaluable asset of trust. Customers and clients rely on businesses to protect their data and maintain confidentiality. A well-handled incident shows responsibility and professionalism, minimizing the reputational damage that can accompany a breach. Conversely, poorly managed incidents can lead to client loss, a tarnished reputation, and years of effort to rebuild relationships. By having an incident response plan in place, small businesses can not only manage crises effectively but also show customers, peers, and regulators that security is a priority. It's a proactive step toward building stability and trust.

While understanding these strategies conceptually is a necessary first step, their effective implementation depends on detailed planning and preparation that is customized to your business's unique needs. This brings us to the role of an incident response policy and playbook.

Playbook vs. Runbook: What's the Difference?

Before diving into planning, it's important to understand the difference between an incident response playbook and a runbook. While they may sound similar, they serve different functions.

Think of a playbook like a football team's strategy guide. It doesn't map out every single step, but it does lay out big-picture plans for different situations, like what to do if the other team is about to score or when to run a trick play. Similarly, an incident response playbook outlines how your team should respond to different "plays," like ransomware or a data breach, ensuring everyone knows their role and works together efficiently.

A runbook, on the other hand, is like an individual player's instructions during a specific play. It provides precise, step-by-step actions—which direction to run, what type of tackle to use, or what to signal. In cybersecurity, a runbook contains technical instructions for IT, detailing exactly how to isolate a compromised system, restore backups, or block malicious activity.

While you'll likely help develop your business's playbook, your IT team or service provider will handle writing the runbook. Both are critical: The playbook provides the overall game plan, while the runbook ensures flawless execution.

It's also important to take your time. An effective incident response playbook isn't something you can throw together overnight. Like disaster preparedness, it requires careful planning, assigned roles, and regular practice drills to ensure your team is ready when an incident strikes.

Building a Playbook Step by Step

Creating a comprehensive incident response policy and playbook requires a lot of planning and attention to detail. It involves developing a clear, actionable framework that outlines exactly how your business will identify, respond to, and recover from security incidents. This process starts by defining roles and responsibilities: who will be involved in managing incidents, what decisions they'll need to make, and how their efforts will be coordinated. Clear communication protocols are equally important, ensuring that everyone knows how to report issues, who to contact, and how updates will be shared during an incident. The process then continues by identifying the potential incidents your company might face. Finally, each incident is outlined with a clear set of steps to resolve it from start to finish.

A playbook should include sections for different types of incidents, including phishing attacks, malware infections, ransomware, data breaches, and failed hard drives. These scenarios should be detailed enough to guide your team through the entire lifecycle of an incident, from detection to resolution, while remaining flexible enough to adapt to unexpected situations. While it's impossible to predict every potential incident, the goal is to anticipate as many scenarios as possible so your team can respond effectively to the unforeseen.

For small and medium-sized businesses, this endeavor is best undertaken with professional guidance. Cybersecurity professionals have valuable expertise when it comes to identifying risks, assessing vulnerabilities, and tailoring solutions to fit unique requirements. They'll collaborate with your IT team to ensure your incident response plan is both thorough and practical. This includes aligning the plan with your available resources, making sure it includes your existing systems, and describing it clearly enough for your team to execute under pressure.

It's important to remember that no two businesses are alike. A one-size-fits-all approach will overlook factors unique to your company. This is why developing a customized plan is so important: It ensures that your business is prepared for the threats most relevant to your environment. If someone offers you a cookie-cutter incident response and disaster recovery plan without first understanding your business and asking the right questions, be cautious!

In the sections ahead, we'll cover the foundational elements of an incident response playbook, including identifying key risks, creating response protocols, and integrating recovery strategies. However, consider these as starting points; building a robust, actionable plan will require additional effort, research, and consultation with experts to refine the details. Ultimately, the goal is to create a framework that equips your team with the tools and confidence to respond effectively to any security incident, and you're not going to accomplish that just by finishing this chapter.

Step 1: Identify Your Threats

The first step in developing an effective incident response (IR) playbook is to identify the key incident scenarios your business is most likely to face. Start by listing out the most common or potentially high-impact threats that could disrupt your systems or expose sensitive data. Think of this exercise as creating a "top threats" list specifically for your company. This commonly includes scenarios like phishing emails, virus infections, and ransomware attacks. You'll want to go beyond the basics by considering the specific risks tied to your industry and even your geographic location. For example, an office in Florida should prepare for potential hurricanes, while an office in California might prioritize planning for wildfires or earthquakes instead.

Don't forget to reflect on past incidents your company has experienced. If you've dealt with a malware outbreak or a social engineering attack before, those scenarios should definitely be included in your playbook. The goal isn't to cover every possible threat, but to build a comprehensive list that focuses on the most likely and most damaging risks.

Step 2: Define Roles, Responsibilities, and Communication Plans

A well-coordinated incident response requires running a tight ship. Everyone needs to know their role, their responsibilities, and who to call when things go sideways. If a cyberattack hits and people don't know what to do, chaos takes over, and that's exactly what hackers count on. While IT plays a critical role, cybersecurity incidents are not exclusively an IT problem. If customer data gets stolen or systems go down, multiple departments—legal, compliance, PR, and executive leadership—need to be in the loop. Legal and compliance teams will handle the messy regulatory side, ensuring you're following breach notification laws to avoid hefty fines. PR and communications control the narrative, keeping customers and partners informed (and preventing a PR disaster). Meanwhile, leadership needs to make the tough calls, such as bringing in external cybersecurity experts, initiating forensic investigations, or even calling in law enforcement.

To keep everything running smoothly, appoint an *incident commander*. This is the person in charge of coordinating the response, keeping everyone on task, and ensuring decisions don't get stuck in endless meetings. This person acts as the crisis manager, directing the team and escalating the response when needed.

Speaking of escalation, not every cyber event is a five-alarm fire. A single phishing email? Probably something IT can handle in-house. A malware outbreak across multiple computers? Now we're looking at an all-hands-on-deck situation. Full-blown ransomware attack or customer data breach? That's when you start making calls to external cybersecurity firms, regulatory agencies, or maybe even the FBI. Mapping out these escalation paths in advance means you won't waste time figuring out who to call while systems are locked down and customers are panicking.

Then there's the communication piece. If handled poorly, this can make a bad situation even worse. Your playbook should spell out who needs to be notified, how they'll be updated, and when those updates should happen. The last thing you want is your own employees to find out about a security breach from a news article or a social media post. Internally, you need secure communication channels, like encrypted messaging apps, an incident response portal, or dedicated email chains. Be careful, though; if your email system is compromised, that may only make things worse!

For external communications, timing and messaging are everything. If customer data is compromised, regulators, customers, and business partners need to be notified, and there are laws governing when and how you should do that. Your playbook should include a clear breakdown of notification requirements based on regulations like HIPAA or state data breach laws. And because drafting statements in the middle of a crisis is a terrible idea, you should have prewritten templates ready to go that include incident reports for stakeholders, press releases for customers, and regulatory disclosures. Having these prepared ahead of time ensures your response remains consistent, is legally sound, and, most importantly, doesn't make the situation worse.

At the end of the day, clarity and speed matter. The faster your team knows what to do, the better chance you have of containing the damage. When an incident happens, you want zero confusion, no scrambling, and no unanswered questions. Everyone should know exactly who to call, what to say, and what to do next.

Step 3: Create Detection and Escalation Procedures

The third step is establishing clear detection and escalation procedures to ensure that threats are identified early and handled appropriately. The faster you can detect suspicious activity and determine its severity, the more effectively you can contain and respond to incidents before they cause widespread damage.

Start by defining how your team will monitor for suspicious behavior. Will you rely on automated tools, such as security information and event management platforms, EDR alerts (refer to the Endpoint Protection chapter), and intrusion detection systems? Define who will monitor these systems and what the process will be for investigating alerts.

Next, clarify what qualifies as an *incident trigger*. This is the threshold that turns suspicious activity into an actual incident requiring a response. For example, if a single employee receives a phishing email, it might be considered a minor event or even ignored as a random occurrence. However, if half of your employees receive the same phishing message within a short time frame, it should be classified as an incident and treated as a coordinated attack. Further escalation may be required if any employee clicks the phishing link.

The playbook should clearly define how to classify incidents by severity—typically low, medium, or high—and outline who is responsible for making that call. This is usually the role of the incident response lead, who assesses the situation, determines the severity level, and triggers the appropriate response.

- **Low severity** incidents are minor and localized, such as a few users receiving suspicious emails. These can often be handled by the IT team without disrupting business operations.
- **Medium severity** involves confirmed threats, like malware detected on a company device. These incidents require containment, remediation, and increased monitoring.
- **High severity** refers to major breaches, like ransomware spreading across servers. These situations demand an immedi-

ate, coordinated response involving leadership beyond the IT department.

By having clear definitions and roles outlined ahead of time, your team can avoid confusion during a crisis and respond faster and more effectively to get your business back on track.

Step 4: Use Visuals for Faster Execution

The next step is to create flowcharts and diagrams to visually map out the response process. In the middle of a crisis, when stress levels are high and every second counts, having visual aids can help teams quickly comprehend what needs to be done without flipping through mind-numbing pages of text. Flowcharts streamline decision-making by presenting clear step-by-step paths based on how the incident unfolds.

For example, in the case of a virus infection, the flowchart might start with detecting the infected system. From there, it could include decision points like "Did containment succeed?" If yes, continue working to remove the virus on the isolated device. If no, escalate by quarantining the network segment and notifying senior IT staff. If the virus spreads beyond one system, the flowchart should indicate when to notify upper management.

Similarly, it should map out different recovery paths based on the level of damage, such as restoring from backups if data is corrupted or restoring the entire server if the damage is severe. By visualizing each step, teams can respond faster, make informed decisions, and reduce the risk of errors that could worsen the situation. Flowcharts are especially useful for training and tabletop exercises, helping everyone understand their role and the overall process before an actual incident occurs.

Step 5: Address Compliance and Legal Requirements

The fifth step in developing your playbook is addressing compliance and legal considerations, particularly if your business operates in a regulated industry, like healthcare, finance, or legal. In these cases, responding to an incident isn't just about containing the threat; it's also about meeting legal and regulatory requirements.

Your playbook should outline the steps to comply with regulations, such as data retention policies, breach notification timelines, and when and how to involve regulatory agencies. For example, a company handling sensitive customer data under HIPAA must notify affected individuals and the proper authorities within a specific timeframe if certain types of data are compromised.

Because these requirements vary depending on your industry and location, you'll need to consult with legal experts or compliance officers when developing this section of your business's playbook. Include specific guidance on which incidents trigger mandatory notifications, such as a breach involving personal data or financial information. Also make sure to document the procedures for preserving logs and evidence in case they are needed for regulatory audits or investigations. This may involve coordinating with legal counsel and external forensics teams to ensure that evidence is collected without tampering.

Step 6: Connect Your Playbook to Technical Runbooks

The final step is to link technical runbooks to each incident scenario. Remember, runbooks serve as detailed, step-by-step instructions for executing specific recovery or containment tasks. Each step in the playbook, such as "Restore the affected servers" during a ransomware scenario, should reference a corresponding runbook that outlines exactly how to perform that task. The runbook will specify the technical details, such as which backup system to restore from, how to verify the integrity of restored data, and how to install security patches before reconnecting servers to the network.

Because resolving complex incidents often involves multiple technical actions, your IT department will need to develop and maintain these runbooks. Attaching runbooks ensures that even in high-pressure situations, team members can follow clearly defined instructions instead of making rushed decisions or improvising. As your infrastructure and tools change and expand, regularly updating your runbooks is important to ensure they remain accurate and effective. By attaching these technical resources to your incident response playbook, you can create a comprehensive plan that marries strategic planning with hands-on execution steps.

One important point to note: Your playbook is only effective if it can be accessed during an incident. Once it's completed, make sure that everyone who might need it knows where to find it and can access it. Distribute the final version of your runbook to all relevant stakeholders, including IT staff, key employees, management, and any external consultants or vendors involved in your response plan. Avoid relying solely on digital storage, as disasters can disrupt access to servers, the internet, and even office computers. Keep several printed copies in secure but accessible locations throughout your business. These physical copies ensure your team can execute the plan even if digital resources are unavailable.

Practice Makes Perfect

Having an IR playbook readily accessible is essential for preparedness, but its true value lies in how effectively it is executed by your team. A well-thought-out plan needs more than just good documentation; it requires hands-on practice so that everyone involved is confident and knows what is expected. This is where tabletop exercises come into play. These activities transform a runbook from a theoretical guide into a practical tool and bridge the gap between writing documentation and real-world execution. By immersing your team in realistic scenarios, you not only test the viability of the plan but also build the competence and coordination needed in your team for high-pressure situations.

Tabletop exercises are an essential component of incident response preparedness, offering a low-stakes environment to test your playbook against hypothetical scenarios. These sessions guide team members through each step of the response process, simulating the pressures and decision-making required in a real incident. The goal is not only to familiarize everyone with their individual roles, but also to establish how those roles work together to achieve a coordinated and effective response.

Beyond role clarification, these exercises also help uncover bottlenecks and inefficiencies before a real crisis hits. These sessions can reveal potential issues, such as communication gaps, unclear escalation paths, or delays caused by dependencies on unavailable resources. By identifying these challenges in advance, the team can make adjustments to streamline the plan.

Practice also ensures that the steps outlined in the playbook are practical, clear, and actionable. Are the instructions easy to follow under pressure, or are there unnecessary complexities that could slow down the response? Regular practice sessions give teams the opportunity to evaluate and fine-tune these processes, bridging the gap between theory and real-world application.

Beyond improving the playbook itself, these exercises reinforce teamwork, building trust and collaboration among participants. This cohesion translates into faster, more confident decision-making, ensuring the business is ready to respond effectively when an actual incident occurs.

This iterative, collaborative approach brings more than just technical benefits; it also fosters a sense of confidence and cohesion among team members. Each practice session strengthens their familiarity with the plan and their roles within it, building trust in both the process and in each other. By testing the playbook in a variety of scenarios, teams can ensure its adaptability to different kinds of incidents, from minor disruptions to major crises. The result is a team that is not only prepared to act swiftly and effectively but also poised to handle the challenges of a real incident with composure and precision. Ultimately, practicing the IR playbook transforms it from a static document into a dynamic tool for resilience, empowering your business to face threats with confidence.

Wrapping Up

Cyber incidents are an eventuality. That makes your incident response and disaster recovery plan a vital lifeline. But having a plan on paper isn't enough. It needs to be practical, current, and tested regularly to truly make a difference when it counts.

Proactive preparation can limit downtime, reduce financial loss, and protect the trust you've built with customers. And while building a comprehensive plan might feel overwhelming, you don't have to tackle it alone. A qualified cybersecurity partner can help craft a strategy that fits your business and keeps you ready to respond.

Next, we'll shift our focus to the human element—empowering your employees to become a powerful first line of defense.

Next Steps

- ☐ Ask if your business has a documented incident response plan or playbook.
- ☐ Identify who would be the main point of contact (incident commander) during a cyber incident.
- ☐ Inquire when the last successful backup restoration test was performed.

Resources:

Watch "Beyond Backups"—a five-minute video that breaks down business continuity planning in plain English and gives you the first questions to kick-start your own plan.

Watch it at smcyber.com/bcp.

PILLAR THREE

PEOPLE

CHAPTER 10

TRAINING EMPLOYEES TO BE CYBER AWARE

If you were designing a security system for a prison, you'd probably think about reinforced doors, surveillance cameras, keycard access, and biometric scanners. Yet, what if you utilized all of that, but none of it mattered?

Several years ago, a cybersecurity firm was hired to test the digital security of a North Dakota correctional facility. You'd think a prison would have robust cybersecurity defenses, right? But instead of cracking passwords or launching a sophisticated cyberattack on their network, the firm used a much simpler method: social engineering.

A woman from the security firm walked into the facility dressed as a health inspector, complete with a clipboard, a fake ID card, and an air of authority. Since real health inspectors often arrive unannounced, the staff didn't think twice about letting her in. She was escorted right past guard stations and given a full tour of restricted areas, all without resistance.

At one point, she was left alone at an office workstation. Within seconds, she plugged in a USB drive containing a backdoor exploit—one that gave her cybersecurity team full access to the facility's internal systems.

No passwords were guessed. No malware was sent. No firewalls were hacked.

The entire facility was compromised simply because the wrong person was trusted.

This is the power of social engineering. It doesn't rely on technical skills—it relies on human nature. Attackers manipulate instincts and trick people into trusting them so they can walk right through the front door. This story isn't unique. In fact, it's the most common type of cybersecurity failure in small businesses: a single employee, without the right training, unknowingly hands an attacker the keys to the kingdom.

That's the failure of traditional cybersecurity training. It focuses on awareness but not behavior. Employees might sit through a security webinar once a year, but when an urgent-looking email arrives in their inbox, instincts take over. Training needs to break those instincts and replace them with a habit of critical thinking.

That's why the People pillar is so critical to your cybersecurity program. Technology can stop a lot, but it can't stop a user from clicking a bad link if they don't know any better. This chapter will help you rethink what effective cybersecurity training looks like and why most businesses are doing it wrong. You'll learn how to build a culture where employees don't just know the rules but think critically and act with caution when something seems off. We'll explore how to:

- help employees recognize the psychological tactics attackers use to manipulate trust;
- encourage healthy skepticism and verification instead of blind compliance; and
- reinforce cybersecurity through relatable, practical activities, not just annual check-the-box trainings.

We'll also cover common mistakes businesses make when rolling out training, creative ways to integrate cybersecurity into daily operations, and why it might make sense to bring in a third-party security awareness training provider.

Because at the end of the day, technology or policies alone won't save your business. Your employees will play just as important a role.

Social Engineering

We've already covered phishing and other scams in previous chapters. You know the tactics: fake emails, fraudulent vendors, and calls from "IT support" that steal credentials.

But why do employees fall for them, even when they've been warned? The answer lies in three core psychological triggers that attackers exploit:

- Urgency – The victim is pressured into acting quickly.
- Authority – The attacker impersonates someone with power.
- Trust – The situation feels familiar and safe.

Every social engineering attack, whether it's a phishing email, a fake IT call, or an intruder walking into your office, exploits at least one of these triggers. The key to stopping these attacks isn't just awareness; it's training employees to override their instincts. Let's break down these three psychological triggers and how your cybersecurity training program can counter them.

Urgency

People are hardwired to react to urgency. When something feels time-sensitive, we respond instinctively—often before thinking critically. Cybercriminals exploit this reflex by fabricating fake emergencies that pressure employees into taking immediate action without verifying the request.

Imagine an employee receives an email from the CEO marked URGENT: "I need you to process this wire transfer immediately! Closing a deal, can't talk, need confirmation in 10 minutes!"

Caught off guard, an employee rushes to comply rather than questioning if the email is real. That's how social engineering works: It manipulates emotions, not systems.

Attackers know that pressure overrides logic. When faced with a deadline or a threat, whether it's a locked account, a late fee, or a delayed business deal, people tend to act first and think later. Employees fear being the reason something goes wrong, and attackers capitalize on that fear by making their requests seem urgent and unavoidable. This is why social engineering scams often frame their messages as time-sensitive: "Your account will be suspended in twenty-four hours," or "Immediate action required to prevent a security breach." Under pressure, even the most cautious employees can be tricked into reacting before verifying.

If employees pause and verify instead of acting impulsively, the attack fails. But without that moment of hesitation, urgency creates the perfect environment for deception to succeed.

Authority

People are conditioned to follow authority, especially in a business setting where employees are expected to comply with requests from executives, IT staff, vendors, and government agencies without hesitation. Cybercriminals exploit this by impersonating figures of power, knowing that most people will obey without question.

A well-crafted email from a "CFO" requesting an urgent wire transfer or a call from "IT support" asking for login credentials is often all it takes. Employees assume that if a request comes from someone senior or official, it must be legitimate, especially if it's urgent. This is why CEO fraud and fake law enforcement or bank calls work so well. No one wants to question their boss or ignore a supposed security warning.

When authority is involved, verification is often skipped. The key to these scams isn't technical sophistication; it's the psychological pressure to comply. And that's exactly what attackers are counting on.

Trust

People naturally trust what looks and sounds familiar. If an email, phone call, or website resembles one they've encountered before, they're far less likely to question it. Cybercriminals exploit this by mimicking trusted brands, vendors, and even internal company contacts to lower defenses and gain access.

An employee who regularly processes payments from a vendor won't think twice when they receive an invoice that looks identical to previous ones: same logo, same format, same email signature. But if that invoice contains updated bank details provided by an attacker, the business could unknowingly send money straight into a scammer's account. Similarly, an IT helpdesk email asking employees to reset their passwords on a cloned version of the company login page can trick even cautious staff into handing over credentials.

This tactic extends beyond emails. Attackers impersonate coworkers, customers, or business partners in phone calls and text messages, using personal details gathered from social media or previous breaches to sound legitimate. If a request appears routine and the sender seems familiar, most people won't question it.

Trust-based attacks don't rely on hacking; they rely on employees filling in the gaps with assumptions. If something looks right, they assume it is right, no questions asked.

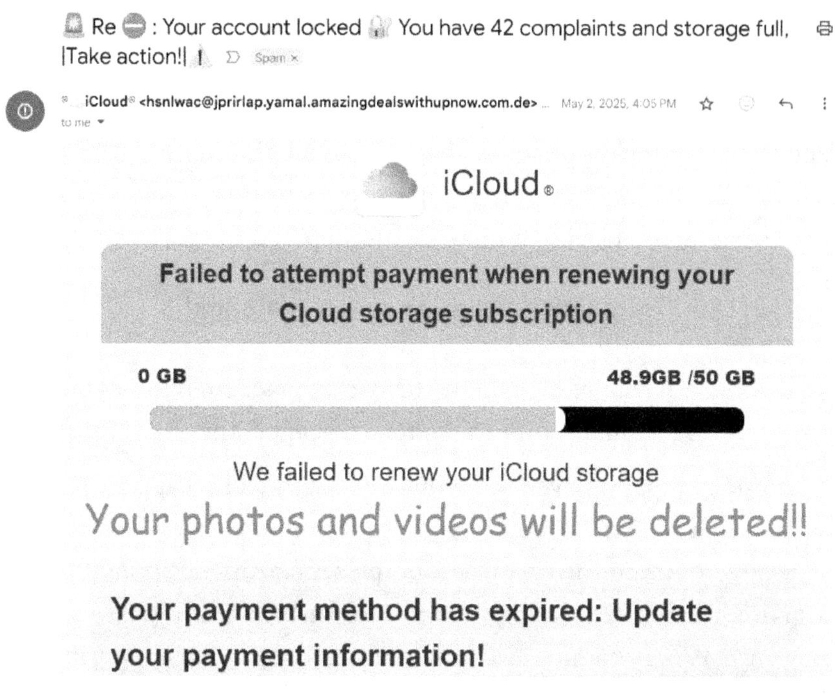

A phishing email received by the author, employing urgency and familiarity

The Biggest Training Mistakes (and How to Fix Them)

Most businesses understand the importance of cybersecurity training, but many fail to implement it in a way that actually protects their organization. They roll out one-size-fits-all courses and host an annual "security awareness day," and then expect employees to instantly become cyber-savvy. But training that doesn't stick is no better than no training at all.

Here are the most common cybersecurity training mistakes and how to avoid them.

Mistake #1: One-and-Done Training Doesn't Work

Too many companies treat cybersecurity training as a check-the-box compliance exercise. Employees sit through an annual PowerPoint presentation or watch a required security video, then promptly forget everything they learned. Meanwhile, cybercriminals are changing their tactics every day.

Threats change constantly, which means security training must be ongoing. Cybercriminals don't send out phishing emails once a year. They send them every day and refine their techniques based on what's working that day. Businesses that rely on one-time training create a false sense of security while leaving employees unprepared for real-world attacks.

Mistake #2: Training is Too Boring or Too Technical

A long, dry lecture on cybersecurity best practices is just as ineffective as no training at all. If employees aren't engaged, they aren't learning.

Some companies overload employees with technical jargon, making cybersecurity seem too complex to understand. Others rely on outdated, passive training formats like long presentations, static slideshows, or hour-long video courses that employees rush through just to get them over with. When training feels irrelevant, employees tune out instead of absorbing the material.

Effective training needs to be interactive, concise, and relevant to the employee's day-to-day tasks. If an employee doesn't see how cybersecurity affects their specific role, they won't retain the information—or worse, they'll ignore security policies altogether.

Mistake #3: Employees Are Afraid to Report Mistakes

One of the most damaging security failures isn't just an employee clicking on a phishing link—it's an employee clicking and not reporting it. Many businesses create a culture of fear around security mistakes, where employees worry that admitting an error will lead to disciplinary action. As a result, they stay silent, giving attackers more time to exploit the breach. A phishing attack that's reported immediately can be stopped before it spreads. But if employees are too afraid to speak up, a single compromised account can escalate into a full-scale breach.

Businesses need to create a no-blame culture in which employees are encouraged to report security incidents without fear of punishment. Cybercriminals rush and pressure people, so the faster the response when mistakes happen, the better.

Mistake #4: Leadership Doesn't Take Cybersecurity Seriously

Employees take their cues from leadership. If managers and executives treat cybersecurity as an IT problem rather than a business priority, employees will do the same.

Some businesses enforce strict cybersecurity policies for lower-level employees, while executives and managers get to bypass them. I've worked with many businesses that force MFA for everyone except the C-suite team. This creates a double standard that weakens security across the entire organization. Senior executives are already among the most attractive targets for attackers because of their elevated access. Skipping basic security measures like MFA doesn't just create a double standard; it hands attackers an even bigger advantage.

For cybersecurity training to be effective, leaders need to set the example. When executives prioritize security, employees will follow suit.

The Emerging Attack Surface of AI

As artificial intelligence continues to evolve, it's opening up new and sophisticated attack surfaces for cybercriminals. One of the most concerning threats comes from AI-powered social engineering, especially through deepfake technology. Deepfakes, which use AI to create convincing audio and video impersonations, can manipulate employees into divulging sensitive information or authorizing fraudulent actions. A deepfake of an executive, for example, could request confidential documents or wire transfers, making it increasingly difficult for employees to discern legitimate requests from malicious ones.

In July 2024, KnowBe4, a prominent cybersecurity firm, hired a North Korea operative who successfully posed as a principal software engineer using deepfake technology. The attacker passed four video interviews, each leveraging AI-generated visuals and audio to change his appearance. This marks a significant turning point, where AI can create convincing digital personas that can easily bypass traditional vetting processes. The use of deepfakes in this case demonstrates how AI can manipulate trust at a scale never seen before, allowing attackers to gain access to sensitive roles without detection. The incident highlights the growing sophistication of social engineering, where AI not only imitates individuals but can also fabricate entire interactions that are virtually indistinguishable from reality.

Beyond deepfakes, AI integrations within enterprise tools, like Microsoft Copilot, present a new and dangerous potential attack surface. These integrations allow AI to interact with a company's systems, making it easier to automate tasks and assist with decision-making. However, if a malicious actor gains access to such tools, they could use them to gather valuable open-source intelligence (OSINT) directly from internal systems. For example, a hacker could use AI prompts like, "What kind of email interactions has the CFO had this last week?" or "Find me all documents containing passwords or financial projections." With this kind of access, an attacker could quickly gather critical information and tailor social engineering attacks more effectively than ever before.

The rapid development of AI means businesses must adapt their security strategies to stay ahead of these emerging threats. As AI continues to advance, attackers will not only use deepfakes and impersonations to deceive employees but will also exploit AI-powered tools to gather and analyze data. To counter these growing risks, businesses need to invest in advanced detection technologies, strengthen access controls for AI-powered tools, and update training programs. As the lines between human and machine continue to blur, organizations must prepare for an era where social engineering tactics are increasingly sophisticated and harder to recognize.

How to Train Your Employees as Defenders

A strong cybersecurity training program doesn't just educate employees; it builds habits that actively prevent attacks. To be effective, training must be structured, practical, and integrated into daily operations. Here's how to develop a program that strengthens your company's security from the inside out.

Before launching a training program, determine where employees currently stand. Some employees may already recognize phishing attempts, while others might not know the basics of password security. A quick security questionnaire and simulated phishing test will help identify weak spots so training can be tailored to actual risks.

Companies can also analyze past incidents, such as security breaches, near-miss phishing attempts, or frequently reset passwords, to spot recurring vulnerabilities. If one department consistently falls for phishing emails, its employees' training should emphasize social engineering awareness.

Once weak points are identified, training should focus on real-world applications rather than generic security briefings. Employees don't need a PowerPoint lecture on cyber threats. They need hands-on experience recognizing and responding to them. One of the most effective ways to achieve this is through phishing simulations: controlled tests where employees receive fake phishing emails. If they click a malicious link or enter credentials, they receive immediate, tailored feedback instead of waiting for an annual security refresher.

Scenario-based training should also be customized for different roles. For example, the finance team needs to recognize fraudulent invoices and payment scams, while customer service teams should focus on social engineering tactics that attackers use to extract customer information. IT staff require more advanced security awareness, including password hygiene, privilege management, and insider threats. The more relevant the training, the more likely employees are to remember and apply what they learn.

Training should also emphasize action, not just awareness. It's not enough for employees to know what a phishing email looks like; they need to know exactly what to do when they see one. A simple but effective approach is the pause-verify-report method: Employees should pause before responding to any unexpected or urgent request, verify it through a secondary channel (such as calling the sender directly), and report anything suspicious to IT or a security lead.

Reporting must be easy. Employees should have a clear, accessible way to flag phishing attempts, whether through a "Report Phishing" button in email clients or a dedicated security channel in the company's productivity software. Just as importantly, they need to understand what happens next. If an employee realizes they fell for a phishing scam, they should know exactly what steps to take: who to notify, how to secure their account, and what to expect in response. When security training is tied directly to real-world decision-making, employees will learn to recognize threats and respond to them instinctively.

For cybersecurity training to be effective, it must become part of daily operations rather than an occasional compliance requirement. The best programs keep security awareness woven into company culture with small, frequent reinforcements instead of long, infrequent training sessions. A quick cybersecurity tip in a weekly team meeting, a short security update in the company newsletter, or a five-minute interactive training module every few weeks helps keep employees engaged without overwhelming them.

Well-timed reminders are also key: Before major holidays, employees should be warned about gift card scams and fake holiday sales, while finance teams should be alerted to wire fraud attempts before the end of each quarter, when scammers are most active. The goal is to keep cybersecurity top of mind in a way that feels natural, not disruptive.

Finally, no training program is complete without measuring its effectiveness and adapting over time. Businesses should track phishing test results, security quiz scores, and incident reports to gauge progress. If employees consistently fall for the same tricks, that's a sign that the training approach needs adjustment.

Training content should be updated regularly to reflect the latest scams and hacker tactics. Employee feedback should also be encouraged. Staff on the front lines often notice security gaps that leadership might overlook. A training program that evolves based on real-world data and employee input stays relevant and ensures employees remain prepared for emerging threats.

Cybersecurity training isn't just about telling employees what to look for—it's about changing how they think and act. The goal is to shift security from a one-time training event into an everyday habit, so employees instinctively question, verify, and report suspicious activity before damage is done. With the right approach, cybersecurity stops being solely an IT concern and becomes what it should be: a company-wide responsibility.

Creating a Security-First Workplace Culture

A company's cybersecurity culture starts at the top. No amount of training, policies, or security tools will protect a business if leadership treats cybersecurity as an afterthought. Employees take their cues from executives and managers. When cybersecurity is modeled, prioritized, and embedded in everyday operations, it becomes an instinct rather than an obligation.

Leaders set the tone by actively demonstrating security awareness in their own work habits. If executives bypass multifactor authentication for convenience or suffer frequent password breaches, employees will assume security isn't that important. Conversely, when leadership practices what they preach by following security protocols, asking employees to verify requests, and calling out security risks, it normalizes security as a company-wide expectation. Employees can't just be told to be security-conscious; they need to see that their leaders actually take security seriously.

Beyond leading by example, leadership must communicate that security is a business priority, not just an IT issue. Cybersecurity isn't just about protecting company data; it is also about protecting customers, maintaining financial stability, and promoting the long-term success of the business. When leadership openly references cybersecurity in company-wide meetings, internal communications, and decision-making processes, it reinforces that security is everyone's responsibility. Employees don't need to be cybersecurity experts, but they do need to understand why their actions matter and how they contribute to the bigger picture.

To reinforce this culture, leadership should embed security expectations into business operations. Security policies should be woven into performance reviews, onboarding processes, and team objectives. Managers should regularly ask employees security-related questions, such as "How would you verify a payment request?" or "What would you do if you accidentally clicked on a suspicious email?" These small, frequent reinforcements ensure cybersecurity remains top of mind, so employees aren't thinking about it only once a year during annual training.

A strong security culture also recognizes and rewards good security habits. Employees who go the extra mile, whether by catching a phishing attempt, identifying a security gap, or properly verifying a suspicious request, should be acknowledged. This doesn't have to be complex or expensive; even a simple "Security Champion of the Month" award, public recognition in a company meeting, or a small gift card can go a long way in reinforcing positive behaviors. When employees see that security is not just a rule but a valued part of their job, they're far more likely to take it seriously.

Finally, leadership must make cybersecurity personally relevant to employees. People are naturally more engaged when they understand how security affects their own lives, not just their company's data. Leaders can promote security by offering resources that extend to personal cybersecurity, such as guides on how to secure home Wi-Fi and protect financial accounts. Some companies even take this a step further by providing security perks. For example, some password manager vendors offer family plans at little additional cost. Businesses that offer this as an employee benefit not only help workers protect their personal accounts but also reinforce the message that security is valuable both in and outside the workplace. A workforce that is trained to protect their own data is far more likely to apply those same security habits to company assets.

At its core, a security-first workplace culture is a leadership-driven initiative. Employees will not prioritize security unless leadership shows that it should be a priority. The most effective cybersecurity programs aren't just built on policies and training—they're built on a culture in which security is expected, practiced, and reinforced from the top down.

Should You Use a Third-Party Security Awareness Training Provider?

Managing a cybersecurity training program in-house can be overwhelming. Developing content, keeping it up to date with threats, running phishing simulations, and tracking employee progress all require time, expertise, and consistency—resources that many small and midsized businesses simply don't have. This is why many companies turn to third-party security awareness training providers that offer structured, scalable training programs designed to keep employees engaged and prepared for real-world attacks.

One of the biggest advantages of outsourcing cybersecurity training is having security experts design the content for you. A third-party provider delivers industry-specific training tailored to real-world threats, ensuring employees aren't just learning abstract security principles but are actively preparing for the exact attacks currently targeting their industry.

Another major benefit is outsourced phishing simulations and attack testing. Running phishing tests internally requires time, planning, and technical expertise, especially to craft realistic attack scenarios that challenge employees without disrupting workflow. Third-party providers handle end-to-end phishing simulation management, designing industry-relevant, up-to-date attack templates and tracking employee responses without burdening internal IT teams. They also provide automated, real-time feedback, helping employees learn from mistakes immediately rather than waiting for an annual review. This ensures that social engineering resilience isn't just trained once but reinforced continuously.

Many third-party platforms also offer interactive learning experiences that go beyond passive training videos. Gamified security training, real-time quizzes, and microlearning modules keep employees engaged and reinforce key lessons without requiring hours of their time. These bite-sized, ongoing reinforcements are far more effective than long, one-time sessions.

For businesses without a dedicated cybersecurity team, outsourcing training also reduces the burden on internal IT staff. Instead of IT managers spending time developing training materials, answering security questions, and running phishing tests manually, a third-party provider automates the process, freeing up IT resources to focus on other security priorities. Additionally, these platforms provide detailed reporting and analytics, helping businesses track training effectiveness, identify employees who need additional guidance, and track overall improvement over time.

For businesses that lack the resources to build a comprehensive, ongoing security awareness program internally, a third-party provider can be a game-changer. By leveraging expert-driven, constantly updated training, businesses ensure that cybersecurity awareness remains effective, engaging, and scalable, protecting employees—and the company—from the latest threats.

Wrapping Up

Building a security-aware workforce goes beyond basic training—it's about shaping habits and instincts. We've looked at why many awareness programs fall short, how social engineering exploits human behavior, and why leadership plays a critical role in setting the tone for a security-first culture.

Whether you handle training internally or bring in outside experts for hands-on simulations, the goal is the same: consistent, practical education that turns your employees into active defenders. Now that your team is equipped to spot and stop everyday threats, it's time to focus on your IT staff and give them the advanced skills they need to lead the charge.

Next Steps

- ☐ Find out when the last company-wide security awareness training or phishing simulation occurred.
- ☐ Confirm if employees know the steps to report a suspicious email or incident.

Resources:

The Employee Cybersecurity Awareness Quiz pinpoints where your team struggles and turns the results into an instant training roadmap. Use it to focus coaching time where it matters most.

Download it here: smcyber.com/satquiz.

CHAPTER 11

TRAINING YOUR IT TEAM FOR SUCCESS

Imagine an entire day without a single IT issue. No slow computers, no forgotten passwords, no security incidents. Now imagine that level of smooth operation for a full week, or even a month. While not every tech problem is a cybersecurity issue, the difference between constant disruptions and seamless operations often comes down to one thing: how well your IT team is trained to think proactively.

This mindset transformation begins with education. A well-informed team doesn't just react to problems. They understand the root causes, especially those linked to cybersecurity threats, and take proactive steps to prevent them. The result? You can worry less about tech glitches and focus on running your business.

No discussion of training would be complete without addressing the backbone of your technology: the IT department. Your IT team plays a critical role as the main line of defense against cyber threats and unexpected technological problems. Unfortunately, many IT teams are stuck in a cycle of reacting to issues as they arise, leaving little time for proactivity or long-term planning. Proper training for your IT team isn't just a nice-to-have; it's a direct investment in your company's future.

This chapter highlights why investing in your IT team's education and development is so important. Training not only helps your team anticipate and address potential risks but also reduces costly mistakes and ensures your business stays compliant with regulations. It builds a culture of learning and adaptability by empowering your staff to embrace new technologies and grow into stronger, more security-aware professionals. As part of the People pillar of your cybersecurity strategy, your IT team needs more than just technical skills; they need the mindset, leadership, and authority to embed security into daily operations. We'll look at how to build a proactive IT culture, why continuous learning matters, and how assigning a dedicated cybersecurity lead can bring focus and accountability to your security efforts.

Why IT Needs Additional Training

In Chapter 3, we compared IT professionals to general doctors: skilled at keeping systems running but not trained to handle specialized security needs. Let's expand on that analogy.

Imagine you walk into your doctor's office for a routine checkup. Your general practitioner can diagnose common issues, prescribe medication, and refer you to a specialist if something serious comes up. But if you suddenly needed emergency heart surgery, you wouldn't expect them to handle it. You'd need a heart surgeon who's spent years studying the heart and how to operate on it. That's the difference between IT and cybersecurity.

Most IT professionals excel at troubleshooting, configuring networks, and ensuring daily operations run smoothly. But cybersecurity requires a unique skill set—one that isn't part of an IT team's core responsibilities. Without specialized training, even the most capable IT professionals may struggle to identify security risks, mitigate threats, or respond effectively to a cyberattack before it causes serious damage.

When IT teams are proactively trained in cybersecurity, the benefits go beyond just reducing security risks. Fewer disruptions mean minimized downtime, smoother operations, and increased employee productivity. Plus, investing in your IT team's development fosters loyalty and retention, keeping your best people engaged and growing with your company. A well-trained team can also help ensure compliance with evolving regulations, reducing the risk of costly fines or legal trouble.

Without proper cybersecurity training, however, IT teams constantly react to fires instead of preventing them. And when businesses rely on a break-fix approach to technology, the costs, both financial and operational, can quickly spiral out of control.

The Cost of Reactive IT

Every minute of downtime and every lost customer is money out the door. Yet, many businesses still play whack-a-mole with technology issues instead of investing in proactive security. The problem? By the time an issue is noticeable, the damage is already done.

Consider a local medical office with two in-house IT techs. For years, their job had been routine: fixing printer jams, handling password resets, and keeping servers online. Instead of proactively managing security risks, they stuck to daily troubleshooting and ignored aging servers, unpatched workstations, and outdated security practices. After all, nothing major had ever gone wrong.

Until a hacker saw the perfect opportunity.

One day, a cybercriminal exploited a weak VPN password and gained access to the network. Thanks to the unpatched, end-of-life servers, it took mere seconds for the attacker to escalate privileges and seize control. But instead of launching an attack right away, the hacker sat quietly in the system—for an entire week.

During that time, backups failed every night with no one noticing. Then, right on schedule, the hacker deployed ransomware. The attack encrypted the office's entire medical records system—along with a week's worth of new patient data that had never been backed up.

The IT team scrambled to respond, but by then, the damage was done. Their options? Pay the $100,000 ransom or lose a full week of patient records.

Many businesses underestimate the cost of downtime. While the IT team tried to recover files, the office couldn't access patient records, schedule new appointments, or process insurance claims. The result was a full week of lost revenue, frustrated patients, and operational chaos.

By a conservative estimate, downtime costs for the average small businesses start at $300 per *minute*. Even a few hours of downtime can be financially devastating. Stretch that over multiple days, and the losses quickly spiral into the hundreds of thousands.

In the end, the medical office's biggest financial hit wasn't the ransom; it was the compounding losses from operational downtime, reputational damage, and long-term business impact. Many patients took their business elsewhere, reluctant to trust a practice that had suffered a major security failure.

This medical office didn't fail because it didn't care about security. It failed because its team thought they had time to fix problems later.

Like many IT teams of small businesses, they focused on "urgent" IT tasks without addressing the bigger issue that their entire security foundation was outdated. Pushing security upgrades back quarter after quarter wasn't just an inconvenience; it created a ticking time bomb.

When companies put off security improvements, the actual cost goes beyond downtime into the steady buildup of risk. The longer security gets delayed, the more outdated systems become, the more compliance gaps widen, and the easier it gets for an attacker to walk in the front door. By the time a breach happens, what could have been a small investment in prevention has snowballed into a crisis.

The companies that survive cyber threats aren't the ones with the biggest budgets. They're the ones that don't delay. They plan, they invest, and they act before they have to.

Building a Smarter IT Team

Many companies view the IT team as problem-solvers rather than risk mitigators. But a truly effective IT team doesn't just troubleshoot security issues after they occur. They actively enforce security measures and integrate them into daily operations. This means moving beyond a passive break-fix approach to one where IT proactively identifies and mitigates risks before they turn into incidents. Instead of waiting for employees to report problems, IT should be conducting security audits, verifying system configurations, and monitoring logs for early signs of malicious activity. The role of IT isn't just to keep technology running but to ensure that every aspect of an organization's technology environment is secured against threats.

An often overlooked aspect of IT's role in cybersecurity is its responsibility to educate and guide employees. Security policies mean nothing if employees don't follow them, and IT plays a crucial role in reinforcing best practices. Every day, IT teams interact with employees who need access restored, permissions adjusted, or account security issues resolved. These are all opportunities for IT to educate users on secure behavior rather than simply granting their requests.

Instead of only resetting a password, IT should investigate why it was compromised. Was it part of a phishing attack? Was it a weak password reused across multiple sites? Does the employee need additional education to avoid making the same mistake again? When IT takes on a mentorship role in security, employees become less of a liability and more of a defense layer against cyber threats.

For IT to be effective in its security role, it must have the authority to enforce changes, not just recommend them. Just like the accounting department approves or denies expenses, and HR has authority over hiring and firing decisions, IT must have the authority to enforce security measures without waiting for executive sign-off. Many security breaches happen not because IT teams failed to spot the risks but because leadership delayed action or blocked technical recommendations. To truly protect the business, IT must be treated as an operational authority in technology decisions, with the autonomy to move quickly when threats arise. Otherwise, IT is trapped in a reactive role, cleaning up after incidents rather than preventing them.

For many small and mid-sized businesses, IT departments are stretched thin. Security is critical, but IT teams rarely have the capacity to monitor every potential risk, analyze every alert, and manually enforce security best practices across the entire organization. AI-powered security tools can serve as a force multiplier by automating security tasks such as vulnerability scanning, log analysis, and anomaly detection. This automation allows IT to focus on strategic priorities rather than reacting to a flood of alerts. Tools like AI-driven EDR can continuously monitor systems and even contain potential threats before they escalate. However, automation alone isn't enough: Expert oversight is still essential to interpret complex threats, fine-tune security tools, and respond decisively when incidents occur. By combining AI automation with expert-led managed detection and response services, supported by an outsourced security operations center, businesses can achieve enterprise-level protection without the need for a large in-house security team.

Businesses that treat IT as a helpdesk will always struggle with security. Those that recognize IT as a critical part of their cybersecurity strategy will dramatically reduce their risk. If IT is sidelined or treated as a purely reactive department, your business will always be one step behind the next threat. To build a truly secure organization, IT must be positioned as both a technical partner and a proactive security enforcer, with the authority to implement changes that protect the business.

Creating a Culture of Continuous Learning

Cybersecurity is a moving target. What worked two years ago is no longer enough to defend against modern threats. Security tools improve, but so do attackers. Cybercriminals are leveraging automation, AI-driven attacks, and social engineering techniques that weren't even on the radar last year. At the same time, regulatory requirements continue to change. The companies that thrive in this environment are the ones that treat security education as an ongoing investment rather than a one-time expense.

Cybersecurity training should be a core part of the IT team's monthly workflow. Learning opportunities don't have to come in the form of long, expensive training bootcamps. Instead, they should be woven into the fabric of IT operations. Short, frequent training sessions, internal discussions about security incidents, and knowledge sharing among IT staff help reinforce best practices in a way that sticks. Encouraging IT teams to stay informed about recent cyberattacks, reviewing how breaches happen, and discussing mitigation strategies keeps security at the forefront of their decision-making process. The more cybersecurity becomes part of IT's everyday responsibilities, the less likely an organization is to fall behind.

While hands-on experience is invaluable, structured learning through certifications and targeted training programs can help IT staff develop expertise in cybersecurity. Many vendors, like Microsoft and Amazon, have security-focused certifications that build on technology you might already be using. Encouraging and supporting IT professionals to pursue these certifications benefits the entire organization. Not only do certified IT staff bring stronger security expertise to their roles, but they are also far more likely to stay with a company that invests in their growth. Many businesses struggle with IT staff retention, but those that actively foster professional development create an environment where IT professionals feel valued and motivated to stay.

Regulatory compliance should be treated as an ongoing learning process, not just an annual audit event. While IT teams play a key role in supporting technical safeguards, true compliance requires participation from leadership, employees, and vendors alike. Staying informed about industry standards like HIPAA, PCI DSS, and security frameworks such as NIST or CIS helps IT teams contribute effectively to a culture of compliance.

A business's security is only as strong as its IT team's knowledge, and failing to invest in continuous learning means falling behind. The question isn't whether your IT team should keep learning—it's whether you can afford for them not to.

Developing a Cybersecurity Lead

Cybersecurity shouldn't be an afterthought within IT. It needs dedicated focus. For small businesses, hiring a full-time chief information security officer (CISO) isn't always realistic, but that doesn't mean security should be left to chance. Instead, businesses should designate a cybersecurity lead. This is a member of the IT team whose primary focus is ensuring the company's security posture remains strong. Having a defined lead to take ownership of security ensures that proactive measures are implemented before they become urgent.

You don't necessarily need to hire a full-time security specialist to strengthen your cybersecurity program. Many times, an existing IT team member can step into a cybersecurity lead role by focusing more deliberately on security efforts. The cybersecurity lead stays embedded within the IT team but takes ownership of integrating security into daily operations, maintaining security policies, coordinating risk assessments, and supporting employee security training. They also serve as a key advisor to business leadership, helping translate cybersecurity risks into business terms and ensuring that security is prioritized across the organization.

The responsibilities of a cybersecurity lead can be divided into daily and long-term objectives. Daily, they monitor security alerts, guide employees on secure practices, oversee access controls, review security logs, and verify that critical patches and updates are applied promptly. Over the longer term, they help develop security policies, oversee compliance tracking, and work with leadership to plan security improvements based on emerging risks. While they may not perform every task themselves, they guide both IT and leadership teams in making security-focused decisions that strengthen the entire business.

For many small businesses, working with external cybersecurity experts is necessary, whether for vulnerability assessments, penetration testing, or incident response planning. The cybersecurity lead serves as the internal point of contact for these third-party vendors, ensuring that recommendations are properly implemented within the organization's IT environment. Without someone driving security efforts internally, important improvements often stall or go unfinished. A cybersecurity lead ensures that security plans turn into real action, not just conversations.

A great starting point for developing a cybersecurity lead is earning an industry-recognized certification. CompTIA Security+ is a widely respected entry-level certification that builds foundational skills in risk management, threat detection, and secure network design. Another excellent option is the Blue Team Level 1 (BTL1) certification from Security Blue Team, which focuses specifically on defensive security operations, threat hunting, and incident response. From there, additional certifications can help deepen prowess in areas such as security operations, incident response, and governance, risk, and compliance (GRC), depending on the needs of the business and the cybersecurity lead's responsibilities. Certifications like Security+ or BTL1 are an excellent starting point, but developing true expertise requires continuous hands-on practice.

Low-Cost Training

Entry-level cybersecurity certifications such as Security+ or BTL1 typically cost less than $500 for the training and exam voucher, making them an affordable investment for most small businesses.

That's where cyber ranges come in—offering immersive environments where employees can apply their skills by defending systems against simulated attacks in a safe, controlled setting. Unlike traditional training focused on theory, these platforms place learners in realistic scenarios that demand quick thinking and decisive action. Many cyber ranges offer structured learning paths and award certificates for completing hands-on exercises, providing clear milestones to track progress. Platforms like TryHackMe, Hack The Box Academy, and CyberDefenders offer beginner-to-intermediate training tracks specifically designed to strengthen security operations and incident response skills.

To supplement structured training, businesses should encourage their cybersecurity lead to stay engaged with the larger security community. Attending industry conferences such as DEF CON, Black Hat, or regional cybersecurity summits helps them stay updated on emerging threats and changing attack techniques. Likewise, taking part in local cybersecurity groups or online discussions allows them to learn from real-world case studies and discuss challenges with peers in the industry.

Small businesses don't need a separate security department, but they do need an IT professional empowered to lead security efforts. The companies that struggle most with cybersecurity are often the ones in which everyone is responsible for security, but no one is leading those efforts. If your business doesn't have a designated cybersecurity lead yet, now is the time to assign one.

Next-Level Thinking: An Introduction to Zero Trust

As your IT capabilities grow and your cybersecurity lead gains experience, you'll start encountering more advanced security strategies. One of the most important of these is the Zero Trust model—a framework that's becoming increasingly relevant for businesses that use cloud services or support remote and hybrid teams. While not every small or mid-sized business is ready to implement it today, understanding the mindset behind it can help shape smarter decisions as your business expands and matures.

You've already taken important steps to secure your business, like enabling multifactor authentication and limiting access using the principle of least privilege. These are your castle walls and moat: strong perimeter defenses designed to keep intruders out.

But in today's environment, perimeter defenses alone aren't enough. Once someone gets inside, traditional models often trust them implicitly. That kind of blind trust can be dangerous. That's where the concept of Zero Trust security comes in. It flips the old Cold War mindset on its head: Instead of "Trust but verify," the new rule is "Never trust, always verify." It assumes that a breach is always possible—or may have already occurred.

This isn't about a single tool or product. Zero Trust is a mindset. Rather than assuming everything inside your network is safe, it treats every request for access as potentially suspicious. Every user, device, or application has to prove it belongs, every time. It's like turning your castle into a series of locked rooms, each with its own guard asking, "Who are you? Why are you here? Should you even be in this room right now?"

This model is especially useful for growing companies that are increasingly cloud-reliant. As your operations expand and your team connects from home offices or client sites, the idea of a fixed "network perimeter" fades. Your company data lives not just on internal servers, but across cloud platforms like QuickBooks Online and Microsoft 365. Zero Trust helps manage all this complexity by applying consistent identity-based rules to control who can access what, regardless of location.

The good news is you don't have to overhaul everything to begin adopting Zero Trust principles. If you use Microsoft 365, you're likely already partway there. Enforcing MFA across all accounts is foundational. So is limiting always-on admin access. Instead of leaving powerful permissions in place permanently, give users elevated access only when they need it. Microsoft 365 also includes built-in identity monitoring through Entra ID, allowing you to see login patterns and flag suspicious activity. Paying attention to who is accessing what, from where, and on what device is part of the continuous verification approach.

Zero Trust is a direction. For smaller businesses, it starts with questions: Where are we too trusting? Where can we tighten access without disrupting work? What activity would signal something's wrong? Even beginning to think this way is a meaningful step forward.

By embedding that "Never trust, always verify" mindset into how you approach users, systems, and data, your cybersecurity lead will be better equipped to protect the business. And when it's time to work with outside professionals, bringing up Zero Trust shows that you're looking beyond today's checkboxes and planning for tomorrow's growth.

Wrapping Up

A resilient cybersecurity posture depends on more than just tools—it relies on a skilled, proactive IT team. When your team is trained to anticipate risks rather than just respond to them, potential crises become manageable incidents. By creating a culture of continuous learning and appointing a clear cybersecurity lead, you give your IT staff the structure and support they need to lead from the front.

Investing in advanced training and embracing strategies like Zero Trust empowers your team to build strong defenses and adapt to an ever-changing threat landscape. Ultimately, your cybersecurity strength lies not in the technology itself, but in the people who manage and maintain it. Preparing and empowering your IT team is one of the most important steps you can take toward long-term business resilience.

Next Steps

- ☐ Approve a training budget and block out time for that first course or certification.
- ☐ Inquire if your IT team is exploring or implementing any Zero Trust security principles.

Resources:

The Fast-Track Cyber Certs Video breaks down the five most valuable security certifications and training programs, so you can recommend the right path for your IT team with confidence.

Watch it here: smcyber.com/certs.

CHAPTER 12

WHEN TO PARTNER WITH PROFESSIONALS

This book isn't designed to transform you into a cybersecurity expert overnight or walk you through every detail of building a comprehensive cybersecurity program. Instead, it's a beginner's guide to give you the clarity and starting points you need to take cybersecurity seriously and build a safer business. Think of it as laying the foundation for a house: While the groundwork is essential, it's only the first step in creating a durable and resilient structure.

Cybersecurity requires continuous attention and a strong balance of people, policies, and technology. And while this book has shown you how to make progress in all three areas, many small businesses eventually reach a point where their internal resources are stretched too thin. Without someone dedicated to leading the charge, it becomes difficult to maintain forward momentum, prioritize effectively, and spot the vulnerabilities that attackers love to exploit.

That's where the Who Not How mindset comes in. Popularized by Dan Sullivan, it's a powerful shift in thinking: Instead of asking, "How do I solve this?", ask, "Who has already solved this before?" Bringing in the right partner isn't a sign of weakness; it's a strategic move that saves time, avoids costly mistakes, and leads to better outcomes.

Partnering with a cybersecurity consultant isn't just about filling knowledge gaps; it also provides structure, accountability, and strategic direction. This external support can empower your internal IT teams, turning good intentions into consistent, focused action. Ultimately, having an expert in your corner helps ensure your cybersecurity remains proactive rather than reactive.

In this chapter, we'll look at how partnering with external experts can strengthen your existing efforts, uncover hidden vulnerabilities, and help you move forward with confidence.

Why Outside Expertise Matters

Cybersecurity isn't just another task on your never-ending to-do list; it's an ongoing battle against constantly changing threats. Yet, for many small business owners, even figuring out where to start can feel overwhelming. When your IT team is already busy keeping daily operations running smoothly, tackling complex cybersecurity issues might seem impossible.

That's where dedicated cybersecurity professionals come in. These specialists aren't just IT generalists; they're trained to anticipate threats, prevent breaches, and spot hidden risks that internal teams, focused on daily demands, often miss. Their specialized expertise and diverse real-world experience allow them to quickly identify vulnerabilities, implement protections, and guide your business toward stronger security without overloading your internal staff.

Importantly, bringing in a cybersecurity consultant doesn't replace your IT team—it strengthens it. A skilled consultant provides structure, accountability, and strategic focus, turning good intentions into consistent action. They offer proven strategies, advanced tools, and cross-industry insights that would take years for an internal team to build on its own. With a cybersecurity expert in your corner, you can move from reactive firefighting to a confident, forward-looking security strategy.

The MSP That Thought it Was Secure—Until it Wasn't

Even IT service providers aren't immune to cyberattacks. That was the hard lesson learned by a local managed service provider (MSP) specializing in agricultural and financial businesses. The MSP advertised cybersecurity services as part of its offering, and a hundred client companies depended on them for IT support, security, and reliable uptime. The MSP believed its own cybersecurity foundation was strong. That is, until a ransomware infection took them down for three days.

It started when a sales team member fell for a phishing email. Within minutes, ransomware spread across the network, encrypting servers, client management systems, and support ticketing tools. Their entire infrastructure had to be rebuilt from backups, every computer had to be wiped and Windows reloaded, and an emergency replacement ticketing system had to be rolled out just to keep fielding client support requests.

After recovery, they called me in for a security assessment and penetration test.

They wanted to validate their security improvements to ensure that what happened once wouldn't happen again. But what I found was alarming: The attack they had suffered could have been much worse, and their improved security measures still had dangerous blind spots. These could have led to another catastrophic breach that wouldn't just take them down but spread to their clients.

One of the first red flags was the use of weak and reused passwords. After running password-cracking techniques, I was able to break into 30 percent of employee accounts, plus a password-protected Excel spreadsheet containing hundreds of client credentials. Even worse, one of the cracked passwords was reused for the domain administrator account, which meant an attacker could have gained full control over every server at their company.

The MSP had also left an end-of-life server on a separate, segmented network, thinking it was secure. While segmentation was a good practice, they had accidentally left an administrator account logged in. That minor oversight gave me the ability to dump the administrator password and reuse it across their other systems, effectively bypassing the network segmentation meant to protect the business. Because the server was end of life, it was vulnerable to the password-dump hack.

Another major weakness was a vulnerable version of their server management platform. By exploiting an unpatched flaw, I was able to create a hidden administrator account that would have allowed an attacker to maintain persistent control over all of their servers, even after the ransomware infection was cleaned up. Additionally, the MSP had enforced multifactor authentication on workstation logins, but they had skipped enabling it on the servers.

But the biggest "oh no" moment came when I looked at their Office 365 security. If an attacker had compromised just one weak employee password, which was synced to Office 365, they could have gained access to a legitimate company email account. From there, the attacker could have launched highly effective phishing attacks from inside the MSP itself, impersonating their employees and exploiting trust to trick clients into handing over credentials or installing more malware. A smart attacker could even have exploited this vulnerability to create dozens of ransomware infections across the business's client base simultaneously, effectively destroying the company. Realizing how easily another attack could have crippled them, with their clients as collateral damage, MSP leadership took immediate action.

I helped them identify and prioritize key security improvements and gave them a roadmap to close these attack vectors before someone else exploited them. But the impact didn't stop there: Their hundred-plus clients also benefited. Using what they had learned from their own security audit, the MSP proactively enforced MFA, improved patching, and enforced better network segmentation across their entire client base. Businesses that could have been next on a hacker's list were now better secured than ever.

This MSP thought they were secure after recovering from an attack, but before a cybersecurity expert assessed their defenses, they had no idea how vulnerable they still were. The same is true for many businesses: Until they bring in a professional to test their security, they don't know what weaknesses are lurking in their networks.

Does Your Business Need a Cybersecurity Consultant?

Deciding when to seek external support depends on the unique risks and challenges your business faces. For many small companies, the tipping point often comes down to industry requirements, operational complexity, or resource limitations.

Certain industries, like finance and healthcare, are subject to strict compliance laws that mandate how sensitive data must be protected. If your business handles regulated information but hasn't conducted a security assessment in the past year (or ever!), you're likely overdue for outside expertise.

Even outside of regulated industries, complexity alone can create hidden risks. Businesses with multiple office locations, for example, often struggle to apply consistent security policies across different teams. Without uniform protections, gaps emerge attackers can exploit. A cybersecurity consultant can help unify your defenses across all sites, ensuring every office remains secure, no matter how small or remote.

Even a strong internal IT team can develop blind spots. Familiar systems, legacy software, and busy schedules can cause vulnerabilities to go unnoticed. If your business is relying on ad hoc security updates or hasn't tested its defenses against modern threats, an external perspective can uncover risks you didn't realize you had.

Ultimately, engaging a cybersecurity consultant isn't about replacing your IT staff; it's about giving them a partner who brings specialized expertise, fresh insights, and proven strategies. This collaboration strengthens your defenses while letting your internal team stay focused on what they do best.

How to Tell if Your IT Provider is Leaving You Exposed

Even with a trusted IT team, hidden vulnerabilities could put your business at risk. Start by asking yourself a few tough questions:

- Has your IT provider conducted a formal cybersecurity risk assessment in the past twelve months, evaluating not just technology, but people and policies, too?

- Are cybersecurity responsibilities clearly assigned to individuals who have specialized cybersecurity training or certifications, rather than just general IT knowledge?
- Has your team practiced a live tabletop exercise and tested full backup restores within the past year to validate your incident response readiness?
- Are critical systems and user accounts routinely audited for weak passwords, unnecessary administrative access, and missing security patches?
- Has your business developed formal cybersecurity policies tailored to your operations, such as an acceptable use policy, password policy, or remote access policy?
- Have your employees taken part in a phishing simulation or hands-on security exercise within the past month?

If you hesitated on any of these questions, you're not alone—and you're not as protected as you might think.

Most businesses don't recognize their cybersecurity gaps until after a breach occurs. A proper risk assessment isn't just a checkbox exercise; it should uncover gaps in not only your technology but also your people and policy layers. Basic vulnerability scans, like those commonly offered by MSPs, aren't enough. A comprehensive assessment evaluates security roles, data handling, user behavior, and technical risks as a whole.

Assigning cybersecurity responsibility also goes deeper than adding "security" to an IT manager's job description. Without specialized training or ongoing education in cybersecurity, even well-meaning teams will fall behind. Real security leaders stay current on emerging threats and industry standards, not just basic IT best practices.

Incident response testing often gets overlooked, yet it's one of the clearest indicators of true readiness. If your team hasn't walked through a live tabletop exercise and proven they can successfully restore backups, you don't have a complete recovery plan. Backup systems are notorious for reporting "green checkmarks" without actually restoring usable data when it matters. Recovery plans should be tested at least annually, with spot checks on critical systems quarterly.

Security policies are another major gap for most businesses. Generic templates won't protect you if they don't match how your company actually operates. Policies like acceptable use and password management need to be understood and followed by your team, not dusty files buried in an onboarding binder.

Employee security awareness must also be treated as an active defense, not a passive formality. Threat actors adapt quickly, and so must your employees. If your team hasn't taken part in a phishing simulation or hands-on security exercise within the past month, they're not truly prepared for the tactics used by today's attackers. Effective security awareness builds instincts through repetition and real-world practice.

Even businesses with strong IT support can be blindsided by unseen vulnerabilities. A cybersecurity assessment isn't about replacing your team—it's about validating that your defenses are complete, current, and ready for real-world threats. Cybercriminals don't care how confident you feel. They care about how well you're actually protected. The only way to stay ahead is to test your defenses before an attacker does.

Wrapping Up

For small businesses, knowing when to bring in outside cybersecurity expertise isn't a weakness—it's a smart move. External consultants offer structure, accountability, and specialized insight that can fill internal gaps and prevent costly mistakes. As we saw in the compromised MSP case study, even seasoned IT providers can overlook critical risks, which makes independent assessments all the more valuable.

Cyber threats are a matter of *when*, not *if*. That's why the next step is action. In the following chapter, we'll walk through twelve high-impact steps you can take right now to strengthen your defenses and reduce your risk.

Next Steps

- ☐ Consider one current cybersecurity challenge: Could a "Who" (an expert) solve this faster/better than "How" (figuring it out internally)?
- ☐ Ask your IT provider: "When was our last comprehensive cybersecurity risk assessment (beyond basic scans)?"

Resources:

Ready to turn knowledge into action but need a personalized roadmap? Schedule a complimentary consultation to get clear, actionable advice on strengthening your business's cybersecurity at smcyber.com/consult.

QUICK WINS AND YOUR NEXT MOVES

CHAPTER 13

QUICK CASTLE FORTIFICATIONS

Throughout this book, we compared your organization to a castle, one that protects valuable treasures like customer data, financial records, and critical operations. But no castle starts out as an impenetrable fortress. Every great stronghold was once a small keep that was gradually reinforced with stronger walls, trained guards, and better defenses.

The good news is that you don't need to build a towering fortress overnight. Cybersecurity is a process, not a finish line. Every step you take strengthens your defenses, making your business harder to attack.

This chapter outlines a "twelve-step program" for strengthening your organization's cybersecurity—twelve practical, high-impact wins you can start implementing right away. While some steps build on policies and best practices we've already discussed, others introduce specific, actionable measures you may not have considered yet. Think of these as your organization's first line of defense: small, strategic improvements that add up to a significant boost in resilience.

Use the checklist below to not only assess where you currently are but also determine which security measures you need to implement next. Refer to this chapter as often as needed to continue fortifying your castle. Even a few well-placed defenses can make your business significantly stronger. Brick by brick, let's build your castle!

Quick Castle Fortifications

1. Use Strong Passphrases

In Chapter 4, we covered how weak passwords are a major security risk, often being the easiest way for attackers to gain access to systems. The goal here is simple: Ensure all accounts in your business are protected by long and unique passwords so attackers can't guess or reuse them.

Start by requiring long, unique passphrases for all accounts, setting a minimum of sixteen characters that combine words, numbers, and symbols. Enforce password uniqueness across different accounts to prevent attackers from exploiting reused credentials. To make this easier for everyone, provide employees with a business-approved password manager that can securely generate, store, and autofill complex passwords.

Take additional steps to ban common and compromised passwords by using tools like Have I Been Pwned's password list or implementing built-in Active Directory password policies. Don't overlook devices either: Change all default passwords immediately on routers, printers, security cameras, and any other business equipment before they go live. Finally, if your industry regulations require it (or if multifactor authentication isn't yet in place), set automatic password expiration policies every six to twelve months to add another layer of protection.

2. Enable Endpoint Protection and Monitoring

Even with strong passwords in place, cyber threats can still slip through other cracks. To protect your business devices, install NGAV or EDR software on all workstations, laptops, and servers, including remote or bring-your-own-device systems that access company data. Configure endpoint security tools to perform automatic scans and enable real-time protection against malware, suspicious activity, and unauthorized changes. Monitoring for unusual behavior is key: Choose solutions that can detect ransomware activity, unauthorized access attempts, and suspicious file modifications before damage occurs.

Be sure to set up alerts for any security incidents and ensure someone is responsible for monitoring them, whether internally or through a managed security provider. Finally, keep your endpoint protection software updated regularly.

3. Automate Software Updates and Patch Management

Outdated software is one of the easiest ways for attackers to breach your business. Start by enabling automatic updates for all operating systems. Also make sure third-party applications such as browsers, office suites, and accounting software receive regular updates. If your business manages multiple devices, use a centralized patch management system to streamline the process.

Network devices like routers, firewalls, Wi-Fi access points, and smart devices also need regular firmware updates to close security holes. Remove or migrate off outdated and unsupported software that no longer receives patches. And for critical systems where updates could disrupt operations, perform test deployments on a small set of machines before rolling them out company-wide.

4. Enable Multifactor Authentication

Even with strong passwords in place, attackers can still gain access through phishing, data breaches, or brute-force attacks. That's why every important account should require at least two forms of authentication: something you know (like a password), something you have (such as a device or hardware key), and something you are (biometrics like fingerprints or facial recognition scans).

Start by requiring MFA on all critical accounts, including email, cloud services, financial systems, remote access platforms, and administrator accounts. Whenever possible, encourage the use of authentication apps like Microsoft or Google Authenticator, or Duo instead of SMS codes, which can be intercepted more easily. For even stronger protection, consider phishing-resistant MFA methods like hardware security keys.

It's important to disable outdated authentication methods, such as legacy email protocols that don't support MFA, to prevent attackers from finding backdoors. VPN and remote desktop connections should also enforce MFA to block unauthorized logins. Lastly, educate employees about MFA fatigue attacks, where repeated prompts are sent to trick users into approving fraudulent access requests.

Hardware Security Keys—The Gold Standard

Hardware security keys stop phishing by confirming you're connected to the real website before approving a login. Unlike codes or apps, fake login pages can't trick them. A hacker would need both your password *and* your physical key to access your account.

5. Maintain Regular Backups

Backups serve as your last line of defense, allowing you to restore data even if ransomware encrypts your systems or a server fails—but only if they're working!

Follow the 3-2-1 backup rule: Keep three copies of your data, stored on two different types of media, with at least one copy located offsite or in the cloud. Automate backups to run daily so that no one has to remember to do it manually. Don't overlook cloud-stored files either! Services like Microsoft 365 and Google Drive don't automatically back up your data—their terms of use place that responsibility on you. Use a third-party backup solution to protect emails, shared drives, and cloud documents.

Protect your backup files with encryption to prevent unauthorized access, and regularly test your backups to ensure they can be restored when needed. To guard against ransomware, maintain at least one backup offline and disconnected from your network.

6. Create an Incident Response Plan

When a cyberattack or security breach strikes, businesses often lose precious time scrambling to react, often making matters worse. A documented and practiced plan ensures your team knows exactly what to do when a crisis hits.

Start by defining clear roles and responsibilities so everyone understands who investigates threats, who contains incidents, and who communicates with leadership, customers, or authorities. Classify incidents by severity to prioritize your response appropriately, and outline specific action plans for common events like ransomware infections, phishing attacks, insider threats, and data breaches.

Document internal and external communication protocols carefully, specifying how and when to notify various parties including stakeholders and law enforcement. Keep backup files and system logs stored separately so attackers can't tamper with evidence. Finally, regularly test your incident response plan with tabletop exercises, allowing your team to rehearse real-world scenarios and refine their approach before a genuine emergency arises.

7. Define Clear Security Policies and Expectations

Even the best security measures can fail if employees aren't clear on what's expected of them. Start by creating a formal cybersecurity policy document that outlines expectations for password management, device security, acceptable technology use, and handling sensitive data. Write the policies in plain language to avoid overwhelming nontechnical employees. Have staff formally acknowledge the policies when they're hired, and again at least annually, to keep security responsibilities top of mind.

Make sure policies also explain the consequences of violations, such as formal warnings or required retraining, so expectations are consistent across the company. If your team includes remote workers, include specific guidelines for securing home networks and personal devices. Finally, review and update all policies regularly to reflect changes in threats, technology, and business operations.

8. Conduct Security Awareness Training

Your employees are the first—and often the best—line of defense against cyber threats. Make security training mandatory for all employees, not just during onboarding, but on a recurring basis. Focus on real-world threats, teaching people how to recognize phishing emails, business email compromise attempts, and common scams. Tailor some of the training based on role. Finance teams, for example, should learn how to spot fake invoices and wire fraud attempts.

Besides recognizing threats, teach employees secure daily habits, such as locking their devices when unattended, handling sensitive information properly, and verifying unusual requests. Keep training engaging by using short, scenario-based videos, quizzes, and occasional gamified challenges to help employees retain what they've learned.

9. Run Phishing Drills and Simulations

Even with training, employees can still fall for phishing attacks, especially as attackers constantly adapt their tactics. Send phishing simulations at least quarterly to all employees, using realistic attack scenarios like fake invoices, urgent credential reset requests, or seemingly normal business communications. Monitor who opens emails, clicks links, or enters information during these simulations, and use the results to fine-tune your training efforts.

Always provide immediate feedback when someone falls for a simulated attack, directing them to a short refresher training that highlights the warning signs they missed. Be sure to include leadership and management teams in these tests to model good security practices. Over time, increase the difficulty of the simulations, moving from obvious scams to more sophisticated spear-phishing techniques.

10. Train the IT Team

IT staff need specialized cybersecurity skills to identify risks proactively, respond to incidents effectively, and develop resilient defenses. Invest in cybersecurity-specific training for your IT team, going beyond general tech skills. Courses like CompTIA Security+ are a sound starting point, but hands-on practice is just as important. Encourage the use of cybersecurity labs and real-world simulations to sharpen detection and response abilities. Make sure IT staff are trained to recognize early signs of security incidents, contain threats quickly, and conduct basic forensic investigations when needed.

To empower your IT team, start by giving them access to basic security logs, system alerts, and monitoring tools they can use to spot suspicious behavior. As their skills grow, they can layer on more advanced capabilities. Require regular exposure to threat intelligence feeds, webinars, and industry briefings to keep their knowledge current. And, most importantly, leadership must visibly support IT security efforts by giving the team the budget, resources, and authority needed to enforce cybersecurity policies across the organization.

11. Build a Reporting Culture

Even with strong technical defenses, fast detection often depends on the people inside your organization. Promote a simple "see something, say something" approach where employees are encouraged to report anything unusual without fear of getting into trouble. Set up an easy way for them to report issues, such as a dedicated email address, phone number, or internal reporting portal.

Focus on helping employees recognize common threat indicators, like phishing emails, unexpected login prompts, or strange system behavior. Reward and recognize employees who report real or potential threats: Small incentives or public shout-outs go a long way toward reinforcing good behavior and making security part of the company culture.

12. Restrict Admin Account Use and Privileged Actions

Controlling who has administrative access is a high-impact security step you can take. Start by separating admin accounts from regular user accounts. IT staff and executives should only use admin privileges when performing administrative tasks and not for everyday work, like checking email. Limit the number of employees with administrative access to the absolute minimum needed to perform critical tasks, and regularly audit those accounts to remove unnecessary privileges.

Where possible, implement just-in-time access controls, granting admin rights temporarily when needed rather than providing permanent high-level access. And make sure all privileged actions are monitored and logged so that unusual behavior, like changes to security settings or unexpected account creation, can be detected quickly.

Wrapping Up

You've now armed your business with crucial 'Quick Castle Fortifications'—tangible, high-impact measures that immediately boost your resilience. From strong passphrases and endpoint monitoring to incident response plans and restricted admin access, each step you've implemented makes your organization a significantly harder target. Remember, cybersecurity is an ongoing journey, not a static destination. The digital threat landscape constantly evolves, making continuous improvement essential. Keep reinforcing your defenses, adapting to new challenges, and building layer upon layer to ensure your business remains secure and thriving. You've started strong; now, let's solidify your commitment to long-term cybersecurity success.

Next Steps

The Castle Fortification Checklist packs twelve quick-win defenses onto one clear page. Print it, tick off your wins, and see at a glance what to tackle next!

Download it here: smcyber.com/castle.

CHAPTER 14

LAUNCHING YOUR JOURNEY

If there's one point you take away from this book, let it be this: Cybersecurity isn't just about stopping attacks. It's about ensuring your organization survives and thrives in a world where digital risks are a part of doing business.

In a world where 60 percent of small businesses cannot recover after a cyberattack, preparation is essential. You've chosen not to wait for a crisis. You've taken action to protect your future, and that puts you ahead of the curve.

Strong cybersecurity is no longer just about defense. It's about building trust, showing resilience, and setting your business apart. Customers notice when businesses take protection seriously. Partners notice. In an uncertain world, strong security becomes a competitive advantage—one that you've already started to earn.

But the work doesn't end here. Cybersecurity is an ongoing discipline, not a one-time software install. But you now have something many businesses don't: a roadmap, a clear understanding, and a head start.

Keep taking small, consistent steps. Strengthen your policies. Train your people. Watch for new risks. Adjust and adapt as your business grows.

You don't have to be perfect—you just have to stay committed.

To continue your journey in cybersecurity, explore further resources, or discuss how to best apply these strategies to your unique business, visit smcyber.com/resources.

The most successful businesses aren't the ones with perfect defenses, nor the ones who spend hundreds of thousands of dollars on the latest cybersecurity subscriptions. They're the ones that invest early, improve continuously, and build trust before it's too late.

You've already started that journey. Now, don't stop, and keep building a future that your competitors will wish they had!

HELP ME!

I'm truly grateful you spent your time with this book. I wrote it to help business owners like you cut through the noise and take control of cybersecurity, without getting overwhelmed.

Now I need your help to make the next version even better!

Your feedback helps me see what worked, what didn't, and how I can improve this book to better serve small business leaders like you.

It only takes a minute, and even a sentence or two makes a big difference.

Leave a quick review here:

smcyber.com/review

Thank you again!!!

— *Seth Bagdanov*

Scan to leave a review!

www.ingramcontent.com/pod-product-compliance
Lightning Source LLC
LaVergne TN
LVHW022038200926
848481LV00003B/719